Flowering Plants of the Galápagos

FLOWERING PLANTS of the Galápagos

Conley K. McMullen

with a foreword by Sir Ghillean Prance
Retired Director of the Royal Botanic Gardens, Kew, UK

COMSTOCK PUBLISHING ASSOCIATES
A Division of CORNELL UNIVERSITY PRESS
Ithaca and London

First published 1999 by Cornell University Press
First printing, Cornell Paperbacks, 1999

Printed in South Korea

Library of Congress Cataloging-in-Publication Data

McMullen, Conley K.
Flowering plants of the Galápagos / by Conley K. McMullen.
p. cm.
Includes bibliographical references (p.) and index.
ISBN 0-8014-3710-5 (cloth).—ISBN 0-8014-8621-1 (pbk.)
1. Botany—Galápagos Islands. 2. Plants Identification.
I. Title.
QK473.32M37 1999
581.9866′5—dc21 99-31321

Cornell University Press strives to use environmentally responsible suppliers and materials to the fullest extent possible in the publishing of its books. Such materials include vegetable-based, low-VOC inks and acid-free papers that are recycled, totally chlorine-free, or partly composed of nonwood fibers. Books that bear the logo of the FSC (Forest Stewardship Council) use paper taken from forests that have been inspected and certified as meeting the highest standards for environmental and social responsibility. For further information, visit our website at www.cornellpress.cornell.edu.

Cloth printing 10 9 8 7 6 5 4 3 2 1
Paperback printing 10 9 8 7 6 5 4 3 2 1

For Peg *and*

Angela Maureen

Consider the lilies,
how they grow;
they neither toil nor spin;
yet I tell you,
even Solomon in all his glory
was not arrayed like one of these.
—Luke 12:27

Contents

Foreword

I feel it is most appropriate that I should contribute a foreword to this book, since it was Charles Darwin's close collaborator and one of my predecessors as director of the Royal Botanic Gardens, Kew, who wrote the first brief flora of the Galápagos Islands in 1847. Since that time a great deal more information about the plants of the Galápagos has been collected by many botanists. Sir Joseph Hooker worked on the specimens collected by Darwin and other early explorers and never had the opportunity of visiting the archipelago. I am fortunate to have made three visits to the Galápagos and, as a consequence, to have spent considerable time trying to identify in the field the plant specimens I encountered.

Much attention has been given to the fascinating evolutionary history of finches and tortoises of the Galápagos Islands, but the evolution by adaptive radiation of various groups of flowering plants, such as *Opuntia* in the cactus family and *Scalesia* in the daisy family, is equally interesting. A good example of this evolution, and one that is easily observable by any visitor to the archipelago, is to be found in the six species and several varieties of *Opuntia*, the prickly pear cactus, which have adapted to local conditions. In some islands, such as South Plaza and Santa Fe, where the giant tortoises occurred, these cacti are treelike, but in others, such as Genovesa, which never had tortoises, they are low-growing. When you get to know the various forms that *Opuntia* has taken, you can usually know which island you are on. Each time I have led tour groups to these islands, I have been able to evoke great interest from the participants by demonstrating the fascinating evolutionary divergence that has taken place in various plant groups. However, in my experience, as in Conley McMullen's, the amateur naturalist or interested layperson has difficulty in using the primary reference book on the Galápagos flora by Wiggins and Porter. That fine book was written mainly for the professional botanist rather than the tourist. What we have lacked is a well-illustrated, user-friendly, and more portable field guide to the Galápagos flora. I am absolutely delighted that this has now been provided by McMullen. Not only has he written a text that is easy to understand; he has also illustrated it abundantly with his own photographs. Instead of the more traditional and often hard-to-use keys for the identification of plant species, the arrangement of the plant descriptions is by features that are easy to observe, such as flower color or the arrangement of leaves on the stem. The plant species that you are most likely to encounter at the various tourist landings in the islands are well covered in this book.

Although it is primarily focused on the needs of the amateur and the short-term visitor to the Galápagos, this book also provides the professional with much new information that cannot be found in any previous publications about the Galápagos. Having studied the pollination of many plants on mainland South America, I have been fascinated by the various observations on plant-animal interactions that are scattered throughout the text. There are even many observations about the famous finches. I thoroughly recommend this book to any traveler to the Galápagos, whether amateur or professional. This will now be an essential component of any Galápagos traveler's kit.

Sir Ghillean Prance

Preface

My association with the Galápagos Islands began in 1983, when, as a graduate student, I traveled to the archipelago to study the reproductive biology of its flowering plants. At that time, the resident botanist at the Charles Darwin Research Station (CDRS) was heavily involved in a campaign to eradicate introduced plants that were competing with the native flora. Despite this project, he and his assistant were eager to share their knowledge of the local plants, accompany me on field excursions, and answer the many questions of a newcomer.

Unfortunately, this enthusiasm was not shared by all members of the community, and plant identification was typically considered best left to the experts. One reason for this attitude was that the primary botanical resource, *Flora of the Galápagos Islands* (Wiggins and Porter 1971), was overwhelming to most readers. It contained a wealth of information on the identification and distribution of Galápagos plants. However, its numerous taxonomic keys and technical descriptions often intimidated the novice. In addition, copies of this work were, and still are, extremely scarce in the archipelago. As a result, a working knowledge of the flora appeared largely inaccessible to all but those with extensive training in the area of plant taxonomy (classification).

In 1984, *Plants of the Galápagos Islands*, written and illustrated by Eileen Schofield, was published in an attempt to familiarize the layperson with several of the more common members of the flora. It included brief descriptions of 87 plants that might be encountered during a visit to the islands. This handy field companion was meant simply as an introduction, so technical terms were kept to a minimum, measurements were not included in the descriptions, and typically only one species from each included genus was described. Thus, the professional botanist and advanced amateur were extremely limited as to the number of species that could be identified. What's more, as with *Flora of the Galápagos Islands*, finding a copy of this book was not an easy task, and it is now out of print.

Consequently, it became increasingly obvious that the time was right for the publication of a field guide that would address the needs of two major audiences: the curious amateur and the professional scientist. Although somewhat of a novelty, this two-pronged approach is not only necessary but natural in a locality such as the Galápagos, where both types of traveler are abundant. The combination of color photographs, concise descriptions, and plant locations will allow the amateur to identify most members of the flora with relative ease. Of special interest will be Appendix 2, which lists those plants most likely to be seen at the major visitor sites in the archipelago. This list can be consulted the evening before an island is visited, giving the traveler an opportunity to become familiar with those plants that will be observed. Descriptions of plant-animal interactions will prove interesting as well. Each of these components will also be appreciated by professional scientists. However, for these individuals many additional details are provided, such as complete species coverage for most genera treated, updated nomenclature, subspecies and variety descriptions, updated distribution records, and much more. These details will make the study of flowering plants easier not only for botanists but for researchers in all disciplines.

Botany is an area of Galápagos natural history that has too long been overshadowed by the archipelago's famed reptiles and birds. One need only to view the numerous films on these islands to realize that the plant life is all but ignored. The user-friendly design of this book will provide most interested people with the opportunity to become excited and familiar with the archipelago's flora and, in turn, to more fully appreciate its importance and relationship to all Galápagos life. This book will prove beneficial to visiting travelers and their naturalist guides, local residents, visiting scientists, CDRS scientists, and Galápagos National Park Service (GNPS) personnel. In other words, it is written for anyone interested in the plant life of these fascinating islands.

Conley K. McMullen

Acknowledgments

No work such as this one can be accomplished without the encouragement, cooperation, and assistance of many individuals and organizations. Each of these deserves to be recognized. I thank my parents, Rev. Conley A. McMullen and Helen K. McMullen, for their love and support through the years. Any successes that I may have had since leaving for college can be credited to the values and faith that they instilled in me as a youth.

A. Clair Mellinger, Norlyn L. Bodkin, and James L. Reveal have been my mentors while serving as major advisers of my B.S., M.S., and Ph.D. programs. The former gave me my first taste of plant taxonomy, while the latter two unselfishly shared their knowledge in this area.

Any research expedition to the Galápagos Islands requires the approval and support of various organizations. The Charles Darwin Foundation (CDF), the Charles Darwin Research Station (CDRS), and the Galápagos National Park Service (GNPS) have always encouraged my studies and provided valuable logistical support. TAME airlines made the cost of travel to the Galápagos Islands more manageable. The majority of my pollination research has been generously supported by the National Geographic Society (Grants 4327–90 and 5014–93). The staffs of these agencies have always performed professionally, and many members have become personal friends. Those whom I have worked with most closely since 1983 include the following officers of the CDF: Craig MacFarland, Juan Black, Alfredo Carrasco, Gonzalo Ceron, Ole Hamann, and Howard Snell; the following directors of the CDRS: Friedmann Koster, Gunther Reck, Daniel Evans, and Chantal Blanton; and the following directors of the GNPS: Miguel Cifuentes, Fausto Cepeda, Humberto Ochoa, Felipe Cruz, Arturo Izurieta, and Eliecer Cruz.

Another group of individuals who have helped enormously with this effort are the resident botanists at the CDRS. Those whom I have worked with directly include Luong Tan Tuoc, Hugo Valdebenito, André Mauchamp, and Alan Tye. Other scientists who have assisted me while in the islands are Basilio Toro, Iván Aldáz, Lenin Prado, and Cruz Marquez.

Over the years I have been accompanied in the field by a number of friends, students, and volunteers. Included among these are Gladys Trávez, Sandra Naranjo, Diane Viderman, Brent Marnell, Jorge Samaniego, Jorge Sotomayor, Patricia Orellana, Jacinto Gordillo, and Scott Shouse. All have proven to be inquisitive, energetic, professional individuals and wonderful field companions.

Many other individuals have shown kindnesses to me during my studies. More than any other person, Marsha Sitnik was responsible for making my first trip to the islands run smoothly. Jacqueline and Tui De Roy have always expressed an interest in my work and provided interesting bits of information on the local plants. Catherine Healy provided me with an opportunity to visit Volcán Alcedo in 1984. David Close assisted with pollen counts. Scott Andrews always supplied film when I was running low. Steve Divine made available for exploration his property in the highlands of Santa Cruz. Linda Cayot, while working at the CDRS, always encouraged my studies. Howard and Heidi Snell are trusted, valued friends, and both have been a wonderful source of support.

Without a doubt, my most enduring friends in the Galápagos have been Godfrey Merlen and Gayle Davis-Merlen. I met these two on my first visit and have now lost track of how many times they've invited me into their home to discuss the local plants, or simply to chat. Godfrey is a naturalist and artist of the first order and has always provided insights on the wildlife of the archipelago. As librarian at the CDRS, Gayle has proved invaluable to me in finding obscure documents dealing with Galápagos botany. In addition, she has endured my many questions via e-mail during the final stages of writing this manuscript.

Obviously, of special importance are the individuals who reviewed the manuscript. Norlyn L. Bodkin, Bryan E. Dutton, Gayle Davis-Merlen, Cynthia H. McMullen, Godfrey Merlen, James L. Reveal, Henk van der Werff, William A. Weber, and two anonymous reviewers made valuable comments on earlier versions of this text. Each of these individuals helped in some way to improve the final product, although the responsibility for any errors is completely mine.

I must also mention my editor at Cornell University Press, Peter J. Prescott. Although he excels in all facets of his job, perhaps his greatest attribute was in making me feel that mine was the only book for which he was responsible. He is a complete professional, and it has simply been a pleasure working with him.

Finally, I thank all of the naturalist guides, starting with Bill Hendricks and Angelica Jahnel, who have waited so patiently for this book. To each of the people mentioned above I am eternally grateful.

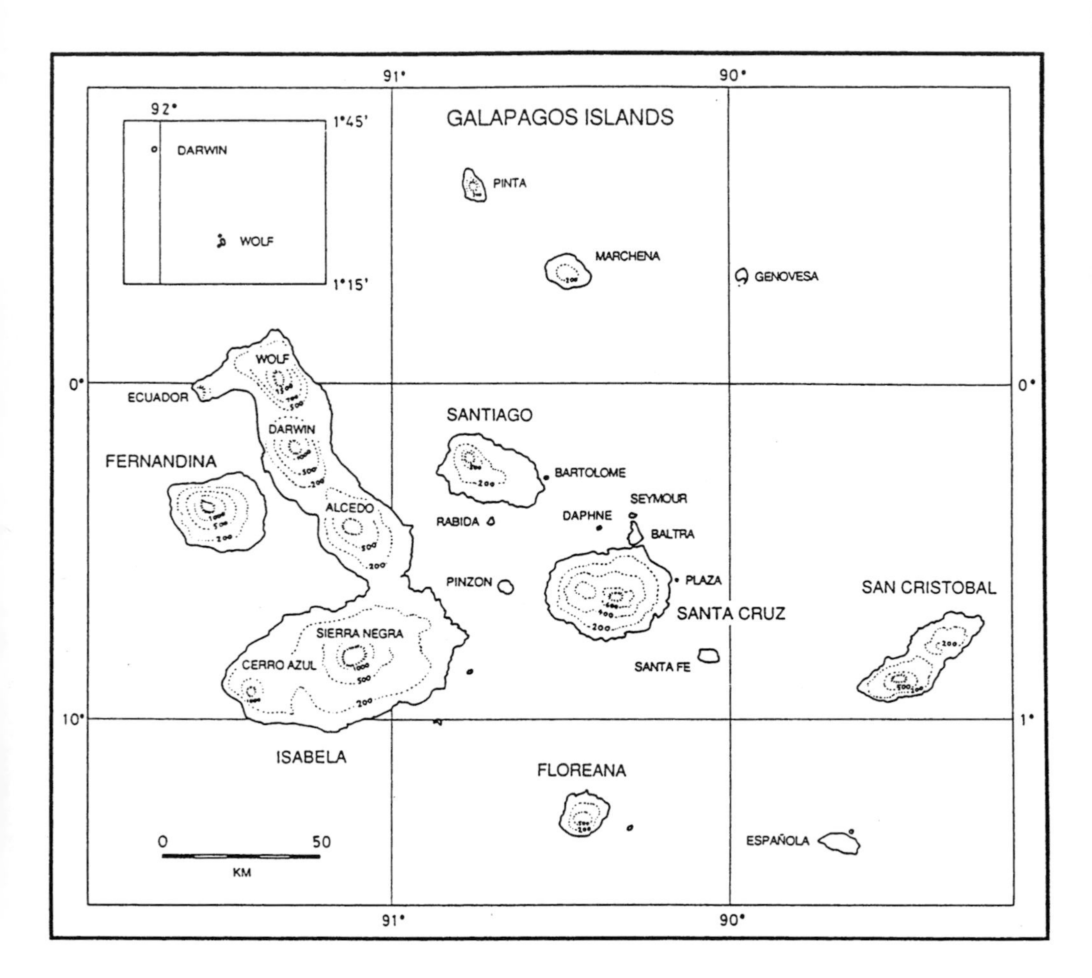

Galápagos Islands, Ecuador. Table 1 lists the English name for each of the islands. Appendix 2 lists selected visitor sites for the archipelago and the plants most likely to be observed at each. Map from *Botanical Research and Management in Galápagos,* ed. J. E. Lawesson et al. in *Monographs in Systematic Botany from the Missouri Botanical Garden* 32 (1990) courtesy of the Missouri Botanical Garden.

How to Use This Guide

This book covers most of the flowering plants likely to be seen by a visitor to the Galápagos. In total, 436 members are discussed, representing 77 families, 192 genera, and 390 species (Appendix 1). Photographs of 266 of these plants are included (76 families, 187 genera, 258 species).

To use the plant key that follows (p. 44), one must first determine and then select the appropriate growth form of the plant being identified. The choices are: tree, shrub, herb, vine, and cactus. In this text, a *tree* is defined as a relatively large, perennial, woody plant that typically has a single main trunk. A *shrub* is a perennial, woody plant with several stems arising from near the base, but no main trunk. An *herb* differs from trees and shrubs in that it lacks a persistent, above-ground, woody stem. In other words, either the entire plant dies at the end of the growing season (annual herb), or the above-ground part dies back at the end of the growing season, leaving only the underground organs, which can renew growth the next year (perennial herb). Occasionally an herb becomes somewhat woody near its base, but it rarely if ever approaches the woodiness of a shrub. A *vine* is an herbaceous or woody plant with a long, flexible stem that trails on the ground or climbs on a support such as another plant or a wall. *Cacti* are grouped in a separate section because whatever their form, shrublike or treelike, they are easily the most recognizable plants in the archipelago.

Some species, such as *Cordia lutea* (yellow cordia) (see Photos 62–63), can have more than one growth form (small tree or shrub). In such cases, the photograph and description of the plant will be listed by its most common form. For *C. lutea* this would be a tree. However, it is cross-referenced with the shrubs, under "See also," since this too is a possibility. As another example, *Desmanthus virgatus* var. *depressus* (slender mimosa) (see Photo 224) sometimes grows as a perennial herb with a woody base, at other times as a small shrub. Its photograph and description are found under the herb category, but it too is cross-referenced under shrubs. As a rule of thumb, if a relatively small plant is not obviously a shrub, then begin by looking under the herb category.

Occasionally one happens upon a plant in which the form, botanically speaking, doesn't fit with what the layperson's intuition suggests. For example, *Musa x paradisiaca* (banana) (see Photo 225) appears to most individuals to be a tree. In reality, it is a treelike perennial herb and possesses no woody above-ground parts. Consequently, it is located under the herb category. A similar situation is encountered with *Furcraea hexapetala* (Cuban hemp) (see Photo 328). This plant is a perennial herb with a short underground stem and large, basally arranged leaves that may be as much as 1 meter in length. The appearance of these leaves often causes one to assume that Cuban hemp is a shrub. However, as with the previous example, the plant has no woody above-ground parts and is listed as an herb.

Each of the five growth form categories, except cactus, is then divided into groups based on how a plant's leaves are typically arranged on its stems. Options include alternate, opposite, whorled, clustered, and basal (see "Illustrated Plant Terms," p. 7). For example, *Tournefortia pubescens* (white-haired tournefortia) (see Photo 131) has leaves that are arranged in an alternate fashion, while *Miconia robinsoniana* (Galápagos miconia) (see Photos 202–3) has opposite leaves, *Mollugo flavescens* (mollugo) (see Photos 324–25) has leaves that are typically whorled, *Lycium minimum* (Galápagos lycium) (see Photo 213) has leaves that are in clusters of two to five, and *Crinum latifolium* (crinum lily) (see Photos 161, 332) has large basal leaves. If a particular plant has dropped its leaves, as often occurs during the cool season, then look for *leaf scars*. These markings on the stems will indicate how the leaves were attached before they dropped off.

Some species exhibit more than one type of leaf attachment. *Porophyllum ruderale* (poreleaf) (see Photos 258–59) has both alternate and opposite leaves. It is described with those plants having alternate leaves, but cross-referenced under those with opposite leaves. A slightly different scenario is encountered with *Bursera graveolens* (incense tree) (see Photos 45–46). Technically, its leaves are alternately arranged, but they often appear clustered at the branch tips. It too is listed with the plants having alternate leaves, but cross-referenced under those with clustered leaves.

All Galápagos members of the cactus family possess clusters of modified leaves known as spines. For this reason, the cactus growth form is not subdivided on the basis of leaf arrangement.

Next, each of the above categories is separated into subsections based on flower color. The choices are *white* (including greenish white, pinkish white, and yellowish white); *yellow or orange* (including greenish yellow, whitish yellow, and reddish orange); *pink, red, or purple* (including purplish pink, reddish pink, orange-red, pinkish red, and purplish red); *blue*; *green* (including yellowish green and whitish green); and *brown* (including greenish brown and yellowish brown). For example, *Scalesia pedunculata* (tree scalesia) (see Photos 40–41) has white flowers, *Opuntia echios* (prickly pear cactus) (see Photo 375–76) has yellow flowers, *Russelia equisetiformis* (coral plant) (see Photo 212) has red flowers, *Commelina diffusa* (dayflower) (see Photo 272) has blue flowers, *May-*

tenus octogona (maytenus) (see Photo 166) has greenish flowers, and *Tournefortia psilostachya* (smooth-stemmed tournefortia) (see Photos 171–72) has brownish flowers.

If each flower of a particular species is composed of more than a single color, then the species is grouped according to the predominant color. *Parkinsonia aculeata* (Jerusalem thorn) (see Photos 64–65) demonstrates this scenario. Its flowers are primarily yellow, but each also has a splotch of reddish orange. Thus, it is grouped with the yellow flowers. One may also encounter a species that produces flowers of a different color on separate individuals. In such cases, the species is listed under the color that is most common in the islands and cross-referenced under the other color or colors. As a case in point, some individuals of *Verbena litoralis* (vervain) (see Photo 322) have purplish flowers, while others have blue or pink flowers. This species is placed under the pink, red, or purple flower category because purplish flowers are the type most often seen. *Lantana camara* (multicolored lantana) (see Photo 201) provides one of the more interesting, if frustrating, challenges for the amateur botanist. Each individual of this species can produce flowers that are yellowish orange, orange, pink, red, reddish orange, and even, on rare occasions, white. It is placed under the yellow to orange group, as these colors are most common. But once again, it is cross-referenced under the other possible colors.

For most species, flower color refers to the plant's petals (corolla) or sepals (calyx). However, some plants produce flowers with neither of these parts. In such cases, the color typically refers to some other flower tissue or to accessory structures such as bracts or glands. For example, *Euphorbia milii* (crown-of-thorns) (see Photo 160) has tiny flowers that lack petals and sepals and that are located in small cuplike structures known as "cyathia." Surrounding each cluster of cyathia are two bright red bracts that simply cannot be overlooked. Thus, this species is placed under the category of plants with pink, red, or purple flowers. *Cyperus anderssonii* (Andersson's sedge) (see Photo 279) also produces flowers without petals or sepals, but beneath each flower is a yellowish brown to reddish brown bract. Consequently, this plant is listed as having "brown flowers."

The reader is advised to keep these "gray areas" in mind when using the key. By doing so, one can identify most of the plants relatively easily. Only rarely does one find a plant that has the potential to cause concern. *Chiococca alba* (milkberry) (see Photo 184) is such a plant. Not only can it take the form of both a shrub and a small tree, but its flowers may be either white or yellow. In fact, the flowers typically start out white but turn yellow with age. But with care, and the knowledge that some cross-referencing may be necessary, the reader will be able to use the key to find a page corresponding to each subsection.

One final hint: Note that not all leaf arrangements and flower colors are listed for each growth form, because not all of these combinations exist in the Galápagos Islands. For example, there are no shrubs with whorled leaves and blue flowers, and there are no herbs with clustered leaves and yellow or orange flowers. This, of course, makes the key a bit easier to use. Once the appropriate subsection is located, the reader simply has to thumb through the photographs until he or she finds one that matches the specimen in question.

Information Items

Each photograph is accompanied by several items of pertinent information, the first being the plant's scientific name. In the case of *Cordia lutea* Lam., *Cordia* represents the genus and *lutea* the specific epithet. Together they form the species name. This system of naming organisms, established by the Swedish botanist Carl Linnaeus (1701–78), is known as "binomial nomenclature." A complete scientific name also includes the author citation. This represents the person or persons who first validly published the plant's name and description. Abbreviations are often used. For example, Lam. stands for Jean Baptiste Pierre Antoine de Monet de Lamarck, the famous French botanist born in 1744. The most common author citation is L., that of Linnaeus. It's important to remember that this terminology is not used to confuse the layperson. Quite the contrary, scientific names often convey interesting and useful information. For example, *lutea* is Latin for "yellow." This refers to the beautiful, bright yellow flowers that characterize this species of *Cordia.*

The scientific name is followed by a common name or names, usually in English and/or Spanish. These are derived from a variety of sources. Many of these names are those used by the local residents of the archipelago. When no such name exists, either one has been borrowed from another locality where the species grows, or a common name has been created based on a dominant characteristic of the plant or a translation of its scientific name. *Cordia lutea* is referred to as *muyuyo* by inhabitants of the islands, but it may also be called "yellow cordia" based on a literal translation of its scientific name. For certain plants, it has been deemed appropriate to use the genus name as the plant's common name. For example, *Blechum pyramidatum* is simply referred to as "blechum."

The common name of each species is followed by the scientific and common names of the family to which it belongs. For *C. lutea* this is Boraginaceae, also known as the Borage family. Following the family name for each species is its geographic range. Endemics are restricted to the Galápagos Islands. Natives occur in the archipelago naturally, as well as in other parts of the world. Introduced weeds and cultivated plants are present as the result of human intervention. Cultivated escapes were brought to the islands for cultivation but have now become established in the wild. The geographic range beyond the Galápagos for each of the natives and exotics is also given. *Cordia lutea* is native to the islands but is also known from mainland Ecuador and Peru.

Next is a list of the major islands where the species is known to occur. Each island, except Isabela, is simply listed by name. Isabela, however, is a special case, as it comprises six large volcanoes that fused together during its formation. These volcanoes are relatively isolated from each other due to distance and/or separation by lava fields, and can almost be thought of as distinct islands themselves (Lawesson et al. 1987). Thus, each volcano that is inhabited by a particular species is listed separately for Isabela. The following abbreviations are used: Volcán Alcedo (A), Volcán Cerro Azul (CA), Volcán Darwin (D), Volcán Ecuador (E), Volcán Sierra Negra (SN), and Volcán Wolf (W). Often the list for a species will end with "Islet(s)." This means that the species is also known from one or more of the smaller islands. Appendix 2 may be consulted if the reader desires more specific information on the location of a particular species.

This appendix lists several popular visitor sites and the plants most likely to be observed at each.

Each species is also assigned to the major ecological zone or *habitat* that it normally inhabits. These are the coastal zone, the arid lowlands, and the moist uplands. Some species occupy more than a single zone. Once again, *Cordia lutea* can be used as an example. It is known from Baltra, Española, Floreana, Genovesa, Isabela (A,CA,D,SN), Marchena, Pinta, Pinzón, Rábida, San Cristóbal, Santa Cruz, Santa Fe, Santiago, Seymour, Wolf, and one or more of the smaller islands, and it typically inhabits the arid lowlands. These lists were compiled using the most current data available to me. They are based on information gathered from a variety of sources, including herbarium collections, scientific publications, photographs, communications from experts in Galápagos botany, and personal records. However, if a particular island is not currently listed for a species, this does not mean that the plant will never be found there. In fact, this may well happen on occasion and simply means that as the islands are explored more completely, the list will become more exact. In addition, as the Galápagos become more traveled, some plants are certain to be transferred from island to island. The visitor should think of this not as a failing of the book but, instead, as an opportunity to be the first to notice a previously unrecorded species on an island.

The species description provides more technical information to complement each photograph. For example, *Cordia lutea* typically takes the form of a small tree or shrub and may reach a height of 8 meters. An important feature of its young branches is that they are covered with fine hairs. It possesses simple leaves that are arranged in an alternate fashion on the stem. In addition, these leaves have an ovate to somewhat roundish shape, are 4 to 10 centimeters long, and have a rough upper surface, a hairy lower surface, and margins that are minutely crenate. Its flowers are arranged in groups known as "cymes." The petals (corolla) are yellow, arranged in the form of a funnel with 5 to 8 lobes, and are 2 to 4 centimeters across at the widest point. Each flower possesses 5 to 8 stamens. The fruits produced by this plant are known as "drupes." They are white in color, roundish in shape, and 8 to 12 millimeters across, and they typically contain 1 to 4 seeds. Initially, the illustrated plant terms (pp. 7–10) and the glossary will prove indispensable when working through this section. However, with practice, one will soon become familiar with these terms, and they will seem second nature.

The last entry for each species includes additional comments of interest, such as the fact that the name *Cordia* honors Euricius (1485–1535) and Valerius (1515–44) Cordus, a father-and-son team of German botanists. In addition, of seven Galápagos species in the genus, *C. lutea* is the easiest to identify because of its fragrant, yellow flowers. Also included are topics such as the plant's benefit to humans and other animals that share the archipelago. The flowers of *C. lutea* are a source of nectar for the Galápagos carpenter bee (*Xylocopa darwini*), and the wood of this tree is a local favorite for making carvings. Special mention is made of those plants currently classified as endangered, vulnerable, or rare by Lawesson et al. (1987). *Endangered* plants are those that may become extinct if the conditions causing their decline are not halted. Plants listed as *vulnerable* may soon become endangered if conditions do not improve. A plant listed as *rare* is one that does not presently fit either of the above two categories but is still at risk.

Information is typically provided on the identification and distribution of all Galápagos members (species, subspecies, varieties) of each genus pictured. In rare instances, only one or a few representatives of a genus are treated. For example, *Alternanthera* includes 13 species, but only three are discussed. In such cases few characters, if any, are available for easy species identification. It should also be emphasized that only Galápagos members of each genus are included in these figures. *Alternanthera* actually contains approximately 200 species worldwide.

The combination of photographs, descriptions, and distributions presented here will make the identification process relatively easy. Let's suppose a visitor encounters a cactus on San Cristóbal. The photographs will allow the observer to make an obvious choice between *Jasminocereus* and *Opuntia*, the two cactus genera that inhabit this island. If it turns out to be the latter genus, then distribution records will indicate that it must be either *O. megasperma* var. *mesophytica* or *O. megasperma* var. *orientalis.* At this point, the descriptions of these two varieties should be consulted. This will show that *O. megasperma* var. *mesophytica* has a trunk to 40 cm across, spines 2.5 to 3.9 cm long, flowers 6 to 8 cm long, and fruits 4 to 6 cm long, while *O. megasperma* var. *orientalis* has thicker trunks (to 60 cm), somewhat longer spines (2.9 to 5.5 cm), longer flowers (8.5 to 11 cm), and longer fruits (6 to 13 cm). This will allow a definitive identification.

Information Sources

The information presented in this book has been derived from a variety of sources, but three of these were relied on extensively and deserve special recognition. The first of these was *Flora of the Galápagos Islands* (Wiggins and Porter 1971). This marvelous book proved invaluable to me with regard to the botanical history and vegetation zones of the islands. Additionally, most of the plant descriptions were based on those in this book. A second major source was *An Updated and Annotated Check List of the Vascular Plants of the Galápagos Islands* (Lawesson et al. 1987). This booklet was useful primarily for the information it provided on species distribution, range, and status (endangered, vulnerable, or rare). Third, most of the descriptions and distributions for members of the genus *Scalesia* came from Eliasson's (1974) detailed study of the genus. These sources were supplemented with updated information by various authors (see "Literature Cited"); conversations with Galápagos scientists, park personnel, and residents; and my own research extending over the past 16 years.

Illustrated Plant Terms

Flower Parts

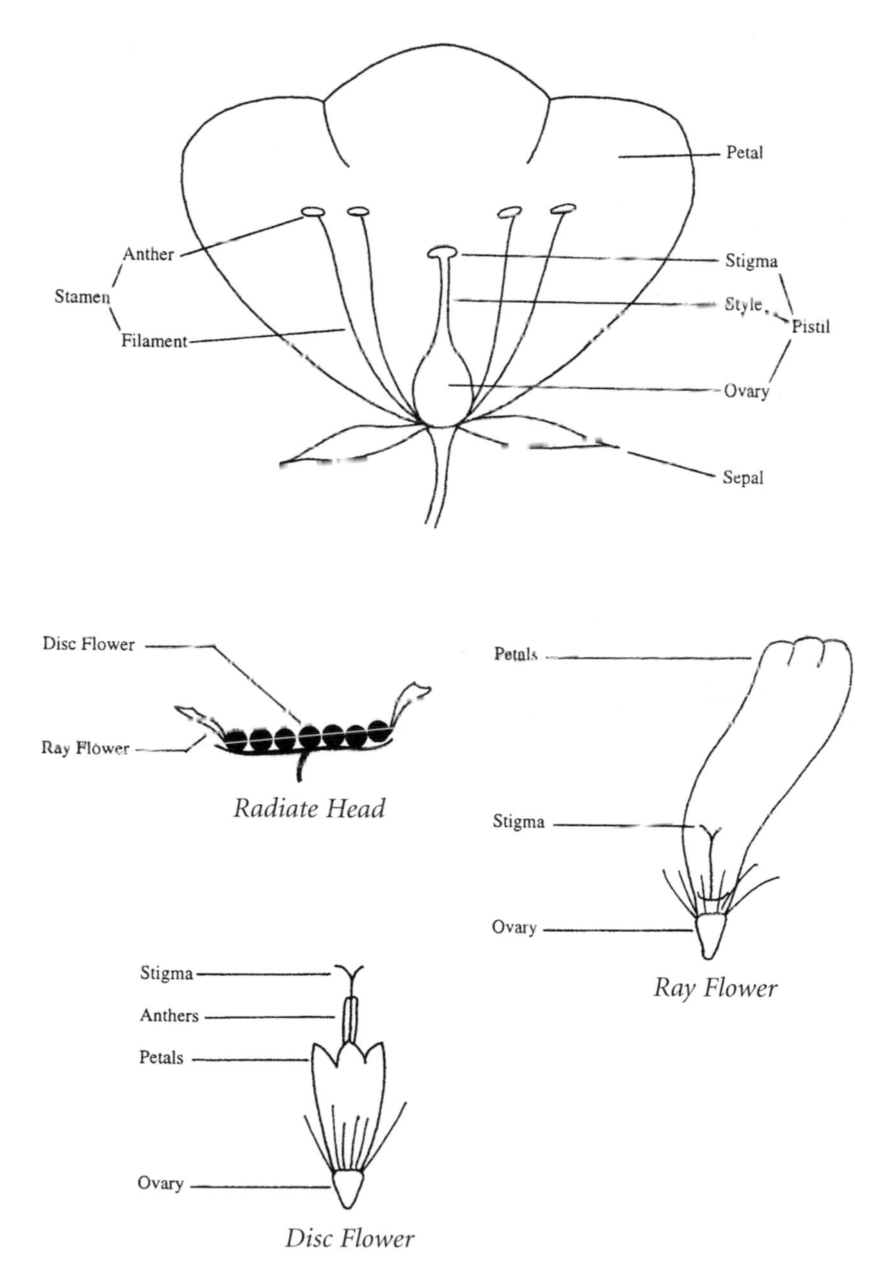

Leaf Arrangements

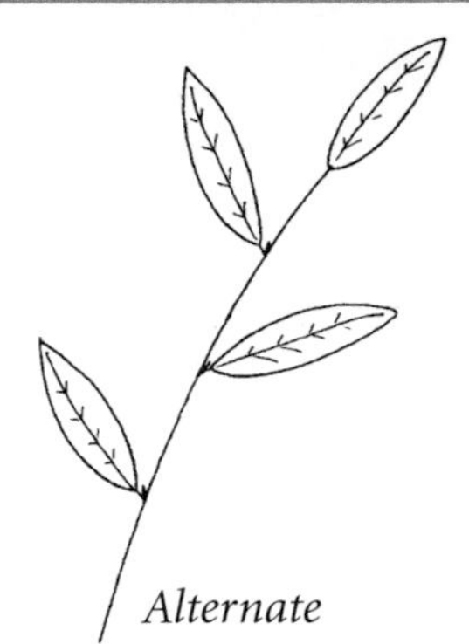

Alternate

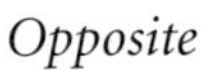

Opposite

Whorled

Clustered

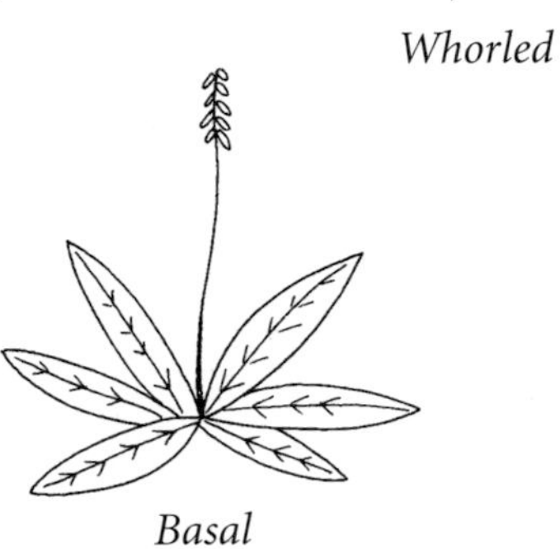

Basal

Leaf Types

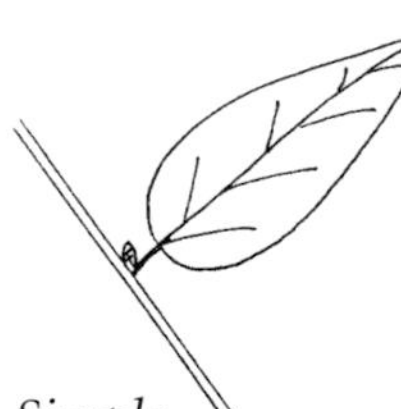

Simple

Even-Pinnately Compound

Odd-Pinnately Compound

Palmately Compound

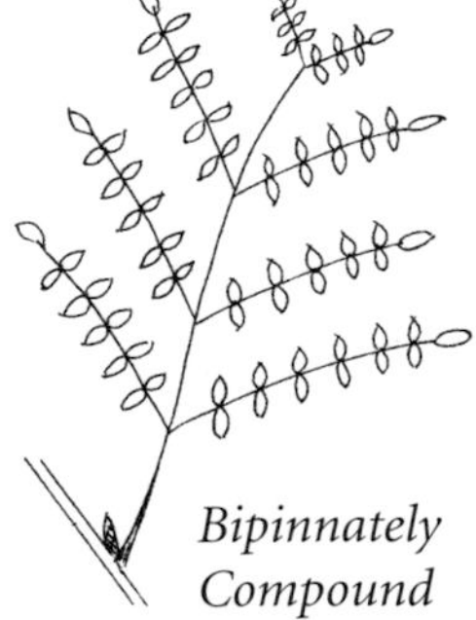

Bipinnately Compound

Leaf Shapes

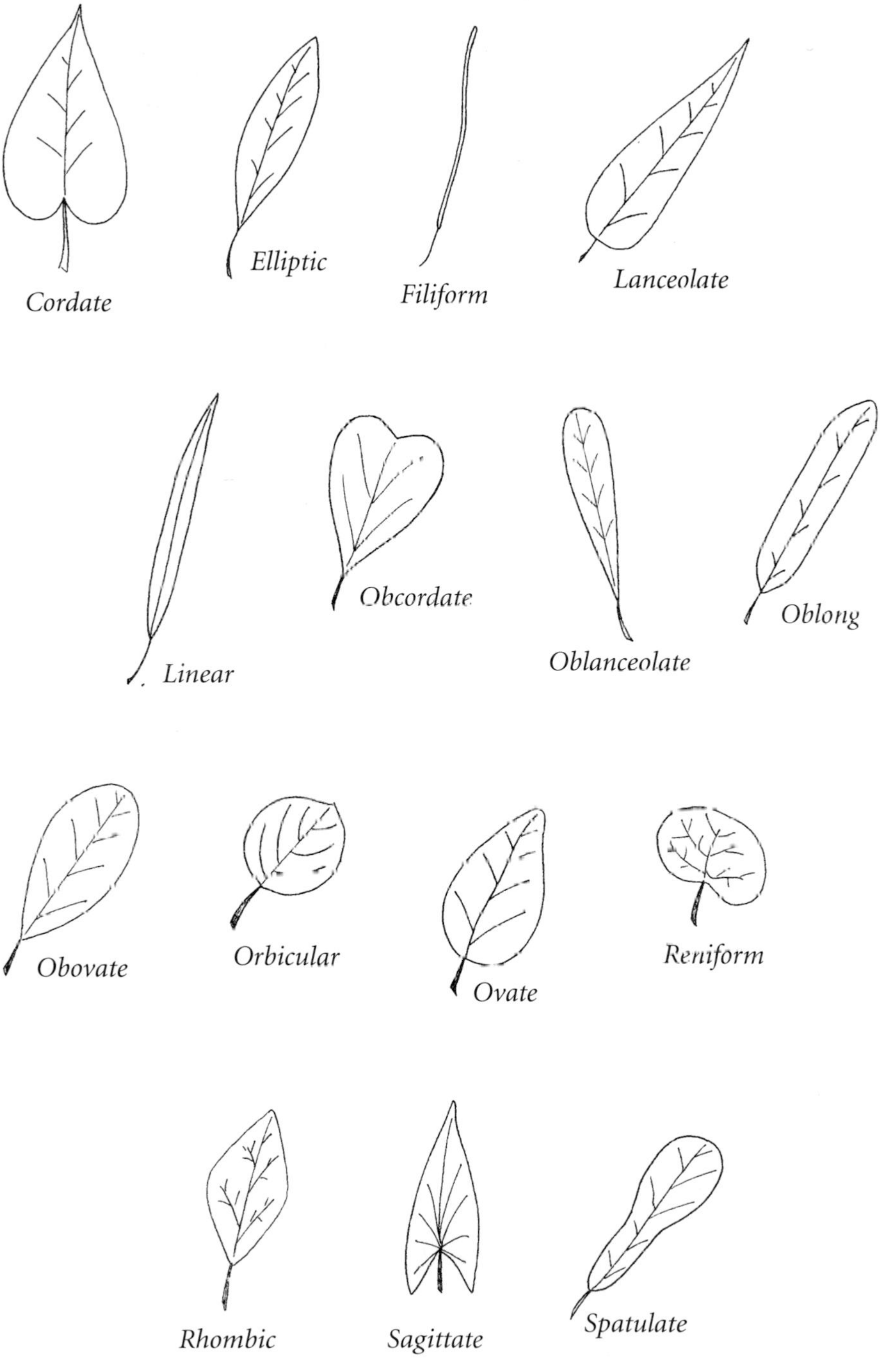

Cordate

Elliptic

Filiform

Lanceolate

Linear

Obcordate

Oblanceolate

Oblong

Obovate

Orbicular

Ovate

Reniform

Rhombic

Sagittate

Spatulate

Leaf Margins

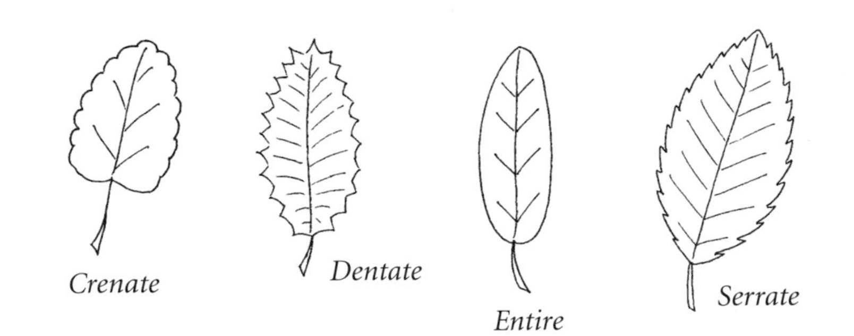

Inflorescence Types

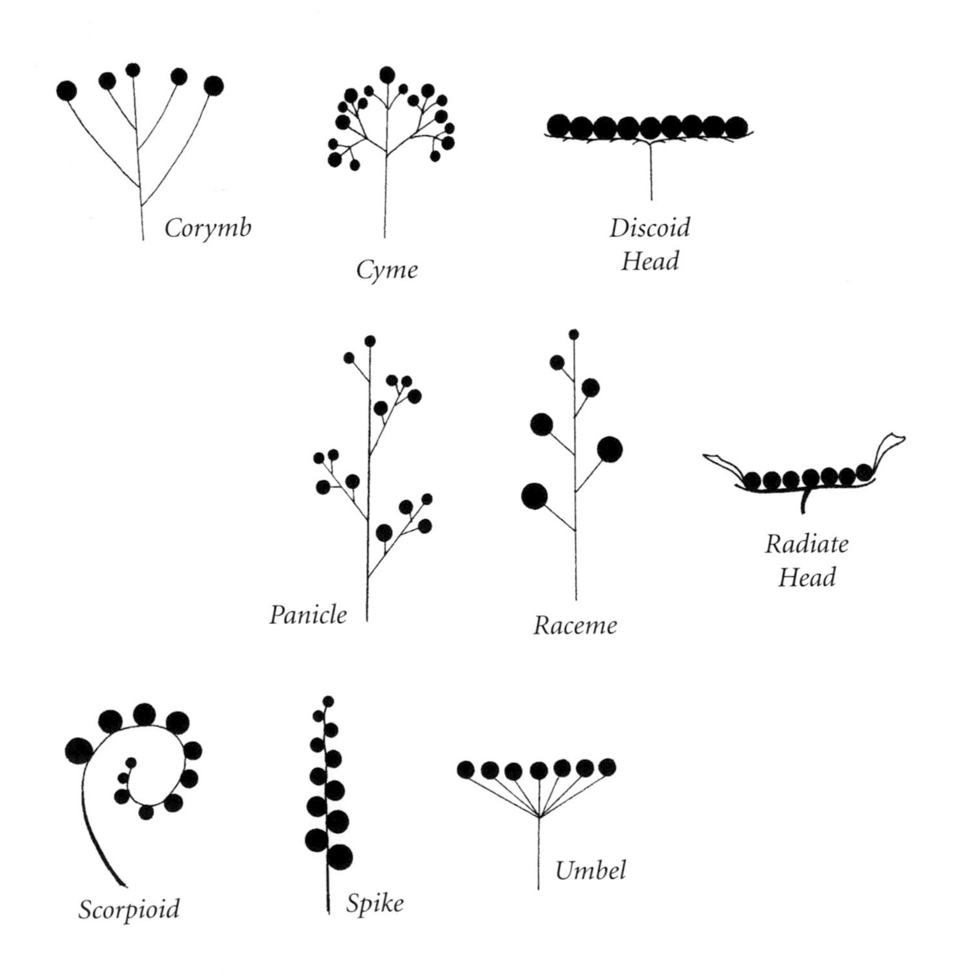

A Brief Introduction to the Galápagos Islands

Discovery

The Galápagos Islands were discovered quite by accident in 1535 by Tomás de Berlanga, Bishop of Panama. Early that year, during a voyage south to Peru, his ship lost the wind and drifted westward for a period of several days. This predicament became life-threatening as water supplies dwindled. However, on 10 March, an island was sighted and plans were made to go ashore and search for fresh water. Seals, iguanas, giant tortoises, and seabirds were encountered in great numbers, but no water (Photo 1). In fact, to survive, the bishop and his men resorted to chewing the pulp of prickly pear cacti for the moisture it contained. This sufficed until they made a landing on a second island and found enough fresh water for the trip back to mainland South America. Tomás assigned no name to this group of islands, although he did relate the experience to his superiors.

In the years since its discovery, this archipelago has been a port of call for travelers as varied as whalers, pirates, fishermen, adventurers, and scientists. Others, such as those looking to start a new life, have attempted more permanent stays. Some of these endeavors have proven wildly successful, while others have met with tragic consequences. Although

1. Marine iguanas (*Amblyrhynchus cristatus*)

not in the scope of this book, these stories make fascinating reading and help one more fully appreciate the rich history of the Galápagos.

Island Names

Eventually this archipelago came to be known as the Galápagos Islands. This name is based on the Spanish word *galápago*, which refers to a type of saddle. It seems that the shells of certain resident tortoises reminded early visitors of these saddles. Indeed, even today some races of Galápagos tortoise are known as *saddlebacks* (Photo 2).

Officially, however, this group of islands is known as *Archipiélago de Colón*, in honor of Christopher Columbus, who, incidentally, never visited the region. And the names of several of the islands are linked with Columbus's famous voyage to the New World. For instance, Pinta and Santa María were named for two of his ships; Fernandina and Isabela for his benefactors, the king and queen of Spain; and Pinzón for Martín Alonzo Pinzón, captain of the Pinta. Over the years, a variety of Spanish and English names have been given to many of these islands (Table 1). This book uses those currently preferred by most inhabitants of the archipelago. As one might guess, the majority of these names are Spanish, as the Galápagos Islands form a province of Ecuador.

Location and Sizes

Comprising 13 major islands and 108 smaller islands (Snell et al. 1995), this Pacific Ocean archipelago straddles the equator approximately 960 kilometers west of mainland Ecuador. Of these islands, Isabela is the largest and highest, with an area of 4,588 square kilometers and a maximum altitude of just over 1,700 meters. It is formed from six volcanoes (Ecuador, Wolf, Darwin, Alcedo, Sierra Negra, and Cerro Azul) that fused together into one giant landmass (Photo 3). Santa Cruz, with an area of 986 square kilometers and an elevation of approximately 860 meters, is second in size and fourth in height. Of course, many of the islands are much smaller (Photo 4), and some are barely

2. Saddleback tortoise (*Geochelone elephantopus* subsp. *hoodensis*)

Table 1. Islands mentioned in the text. The preferred name is in bold type.

Spanish name(s)	*English name*	*Spanish name(s)*	*English name*
Baltra	South Seymour	**Marchena**	Bindloe
Bartolomé	Bartholomew	**Mosquera**	Seal Island
Corona del Diablo, Onslow	Devil's Crown	**Pinta**	Abingdon
		Pinzón	Duncan
Daphne	Daphne Major	**Plaza Sur**	South Plaza
Darwin	Culpepper	**Rábida**	Jervis
Española	Hood	**San Cristóbal**	Chatham
Edén	Eden	**Santa Cruz**	Indefatigable
Fernandina	Narborough	**Santa Fe**	Barrington
Floreana, Santa María	Charles	**Santiago**, San Salvador	James
		Seymour	North Seymour
Gardner	Gardner-near-Hood	**Sombrero Chino**	Chinese Hat
Genovesa	Tower	**Tortuga**	Brattle
Isabela	Albemarle	**Wolf**	Wenman

3. Volcán Darwin and Volcán Wolf as seen from Volcán Alcedo (Isabela)

4. Daphne and Daphne Minor as seen from Seymour, Galápagos sea lion (*Zalophus californianus* subsp. *wollebaeki*)

visible above the waves. In all, the total land area of the archipelago approaches 7,900 square kilometers.

Geology

Technically, the Galápagos are *oceanic islands*: They were produced by volcanic activity and have never been part of the mainland. Mere specks on a map, they are actually the tops of mountains that rose from the ocean floor and broke the surface less than 5 million years ago. The youngest islands are located in the northern and western parts of the archipelago, while the oldest are found to the southeast. This is because there is a stationary volcano-producing area in the earth's mantle near the western edge of the archipelago. This area is referred to as a *hot spot*, and it results in the formation of volcanoes on the surface of the tectonic plate above it. The plate involved, actually a giant piece of the earth's crust, is known as the Nazca Plate. It is slowly moving to the southeast at approximately 7 centimeters per year, carrying these volcanic islands along with it. As the first-formed islands moved eastward, other volcanoes broke through the plate to replace them. Thus, not only are the western islands younger, but they are also more active volcanically. In fact, several relatively recent eruptions have occurred on Fernandina (1978, 1984, 1988, 1991, 1995) and Isabela (1979, 1982). Ultimately, those islands now considered part of the archipelago will be forced under the South American continent. Before that time, however, others will no doubt have replaced them.

Galápagos volcanoes are typically somewhat shield-shaped and composed of basaltic lava. Each volcano comprising Fernandina and Isabela, which are of relatively recent origin, has a giant craterlike depression in the center known as a *caldera*. Volcán Alcedo's caldera on Isabela, famous for its tortoise population, is approximately 6 kilometers across (Photo 5). The older islands have been exposed to the effects of erosion for much

5. Caldera of Volcán Alcedo (Isabela)

longer periods of time and are somewhat less dramatic. As lava begins to cool after an eruption, it can take one of two basic forms. The first has a smooth, somewhat ropy appearance and is known as *pahoehoe* (Photo 6). The other, referred to as *aa*, is rough and broken and makes for extremely difficult walking (Photo 7).

Climate

The Galápagos Islands, which are extremely arid compared to most tropical archipelagos, experience two fairly distinct seasons each year. The warm season, typically January through May, is caused by warm ocean currents sweeping southward from the direction of Panama. These currents cause both sea and air temperatures around the islands to rise. During this time the skies are normally clear, with occasional heavy showers (Photo 8). This is when the lowland vegetation reaches its peak.

The cool season lingers from June to December. The weather at this time is created by the Humboldt (Peru) Current, which brings cold water north from the Antarctic along the west coast of South America and then westward through the archipelago. This results in cooler air temperatures. Usually the skies are overcast, with little precipitation

6. Pahoehoe lava

7. Aa lava

8. (above) Typical warm season weather on Santa Cruz

9. (right) Galápagos hawk (*Buteo galapagoensis*) surrounded by garúa (Pinta)

in the lowlands. However, many parts of the highlands are constantly wet at this time due to a mixture of light rain and mist, known locally as *garúa* (Photo 9).

It must be remembered that these statements on the archipelago's climate are generalizations. Height, exposure, angle of slope, and overall size determine the amount of rain that an individual island will actually receive (Hamann 1981). For example, during the cool season the prevailing winds are from the southeast, so the southern and eastern slopes on the higher islands tend to be wet, while the northern slopes are rather arid. Lower islands receive little if any moisture at this time (Photo 10).

A discussion on the climate of the Galápagos Islands is not complete without a mention of the phenomenon known as *El Niño*. This event, named for the Christ child, involves the movement of warm waters from the western part of the Pacific through the archipelago. This increase in ocean temperature often results in an unusually large amount of rainfall. The effect on the archipelago's vegetation is varied. Many of the succulents, such as *Opuntia echios* (prickly pear cactus), receive too much water. This results in the weakening and death of many of these cacti and a consequent decline in their

10. Typical cool season weather on Bartolomé and Santiago

population. Conversely, these rains allow many other species to thrive as at no other time. Plants such as *Passiflora foetida* (running pop), *Mormordica charantia* (bitter melon), and *Merremia aegyptica* (hairy merremia) (see Photo 340) literally cover the ground and climb over any plants they encounter. Unfortunately, this often exacerbates the cacti's problems by causing the weaker individuals to topple over. The changes in the archipelago's vegetation from year to year can be truly remarkable. As a consequence, it takes many visits to the islands over several years before one can begin to intimately know the flora and its dynamics.

Early Botanical Studies

The Galápagos Islands remained botanically unexplored for almost 260 years after their discovery. But in February 1795, Archibald Menzies, while acting in his capacity as naturalist on HMS *Discovery*, collected what are thought to be the first plant specimens taken from the archipelago (Porter 1980a). Although few in number, these collections from Isabela are nonetheless important from a historical perspective.

In 1825, two British naturalists, David Douglas and John Scouler, briefly visited the island of Santiago and collected several plants. Unfortunately, preserving and transporting scientific specimens was more difficult in those days, and only a few of the plants made it back to England. Despite this problem, Scouler's efforts were later honored when *Cordia scouleri* (Scouler's cordia) and *Croton scouleri* (Galápagos croton) were named. This practice of naming plants for individuals becomes apparent as one learns the scientific names of Galápagos organisms and the history of these islands. For example, James Macrae also visited the archipelago in 1825 and made 41 plant collections on Isabela. The genus *Macraea* honors this contribution.

On 15 September 1835, Charles Darwin began a five-week visit, the consequences of which would ultimately bring lasting fame to the Galápagos Islands. As any student of biology knows, while serving as naturalist on HMS *Beagle*, Darwin made certain critical observations that would eventually contribute to his theory of evolution by natural selection. He noted, for example, the interisland variation that exists among the shells of the giant tortoises that inhabit the archipelago. Also of some interest was the fact that Galápagos finch species have very different beak shapes. What most people don't know, however, is that Darwin collected a sample of every plant that he observed flowering on San Cristóbal, Floreana, Isabela, and Santiago. He appears to have been somewhat underwhelmed by the experience, going so far as to describe the flora of San Cristóbal as composed of "wretched-looking little weeds" (Darwin 1845). Nonetheless, his collection amounted to 173 different plants, many of which were new to science (Porter 1980b).

Specimens collected by each of the above-mentioned explorers (except Menzies) and gathered by a few other early visitors (Hugh Cuming, Thomas Edmonston, Abel Du Petit-Thouars, John Goodridge) eventually were examined by Sir Joseph Dalton Hooker of the Royal Botanic Gardens, Kew. These formed the basis of *An enumeration of the plants of the Galápagos Archipelago; with descriptions of those which are new* (Hooker 1847), which represents the first attempt to describe the flora of these islands.

No major collecting trip occurred again until 1852, when N. J. Andersson, a Swedish scientist, spent 10 days in the archipelago. His visit is notable in that he was the first to botanize the island of Santa Cruz. The fact that it took so long for this island to be explored is somewhat surprising, since Santa Cruz is the second largest island and lies in the middle of the archipelago. In any case, Andersson made 325 different collections (each collection may include several duplicate specimens). Fifty of these represented new species.

In the years following Andersson's visit, several other scientists made brief excursions to the islands. These resulted in the discovery of only a few new species. The next extended stay took place in 1891, when George Baur spent three months in the archipelago. He visited 13 islands and made 385 collections of vascular plants, several of which were not previously known.

R. E. Snodgrass and Edmund Heller, in collaboration with Stanford University, spent six months in the Galápagos beginning in December 1898. They visited 16 islands and gathered some 949 collections of vascular plants. Wiggins and Porter (1971) note that these two scientists appear to have been the first to botanize Baltra, Darwin, Fernandina, Wolf, and perhaps Seymour. Less than four years later, the state of floristic knowledge in the archipelago was summed up in B. L. Robinson's *Flora of the Galápagos Islands* (Robinson 1902).

Twentieth-century botanical studies began in earnest when an expedition from the California Academy of Sciences toured the islands for just over a year beginning in September 1905. The botanist on this voyage, Alban Stewart, managed to visit most of the islands and collect over 3,000 numbers. These specimens formed the basis of *A botanical survey of the Galápagos Islands* (Stewart 1911). In this publication, 615 species, subspecies, and varieties of vascular plants were catalogued. Only 10 of these were new to science. As a note of interest, Academy Bay, located on the south side of Santa Cruz, was named for the expedition's ship *Academy*.

Henry K. Svenson, representing the Brooklyn Botanic Garden, spent two weeks in the Galápagos in 1930. He collected some 300 numbered specimens of vascular plants from the islands of Baltra, Floreana, Genovesa, and Santa Cruz. Two were described as new species. Two years later, John Thomas Howell of the California Academy of Sciences spent two months studying the flora and collecting specimens. This research resulted in numerous manuscripts describing various plant taxa; those on the cactus family were perhaps the most noteworthy.

As one might expect, floristic surveys of the Galápagos Islands were not a high scientific priority during World War II. However, studies began anew in the 1950s, with the arrival of Scandinavian botanists Folke Fagerlind and Gunnar Harling. Up to this point, scientific research in the Galápagos had largely been limited to brief visits by wealthy individuals eager to experience the unique character of these islands, or by relatively small expeditions sponsored by universities or learned societies. Typically, these studies involved collecting as many specimens as possible in the time available. Upon returning home, the investigators could examine their collections at leisure. Difficulties connected with long-term investigations in the field had not yet been surmounted.

The presence of a permanent scientific entity in the archipelago was guaranteed in 1959 with the establishment of the Charles Darwin Foundation (CDF) for the Galápagos Islands. This event occurred, by no coincidence, on the 100th anniversary of the publication of Darwin's *Origin of Species* (Darwin 1859). It was followed by the inauguration of the Charles Darwin Research Station (CDRS) in 1964, which opened a whole new world of possibilities for scientific investigations. In fact, that very year, during the Galápagos International Scientific Project, Ira L. Wiggins collected almost 1,000 numbered plant specimens from Santa Cruz. Two years later, Paul A. Colinvaux collected over 400 specimens while studying the historical pollen record of the Galápagos Islands. And in 1967, Ira L. Wiggins and Duncan M. Porter made 571 collections of vascular plants. This was the last major study before the establishment of the Galápagos National Park Service (GNPS) in 1968. This event, which resulted in approximately 97% of the archipelago being declared a national park, greatly increased the chances that its unique flora would be available for future generations to enjoy.

The previously mentioned studies, and many others of a more limited nature, led to the publication of *Flora of the Galápagos Islands* (Wiggins and Porter 1971). This impressive work included a total of 702 plants, 612 of which were flowering plants. Since its appearance, botanical studies have accelerated, and scientists too numerous to mention by name have added to our overall knowledge of the plants inhabiting this archipelago. However, certain individuals have continued to contribute to Galápagos botany over the years and merit special mention. These botanists and a sampling of their contributions include Henning Adsersen (conservation, floristics, taxonomy of *Lecocarpus* and *Scalesia*), Uno Eliasson (floristics, taxonomy of *Macraea* and *Scalesia*), Ole Hamann (conservation, ecology, floristics, taxonomy of *Scalesia*), Jonas Lawesson (ecology, floristics, taxonomy of *Darwiniothamnus* and *Passiflora*), Duncan M. Porter (biogeography, botanical history, floristics), Charles M. Rick (plant-animal relationships, pollination biology, taxonomy of *Lycopersicon*), Eileen Schofield (bibliographical studies, floristics), Henk van der Werff (conservation, ecology, floristics), and William A. Weber (floristics, taxonomy of lichens and mosses).

Another select group of individuals has directed Galápagos plant studies while serving as resident botanists at the CDRS. These include Luong Tan Tuoc, Jonas Lawesson, Andrew Schmidt, Hugo Valdebenito, André Mauchamp, and Alan Tye. Their contributions to the conservation of this archipelago's flora cannot be overstated. Jacinto Gordillo, acting as a representative of the CDRS, has assisted innumerable botanists with their field studies, sharing his knowledge of the local plants. He even discovered a new species of *Scalesia* that was subsequently named in his honor. Jorge Sotomayor, also a representative of the CDRS, has worked for several years on San Cristóbal to save *Calandrinia galapagosa* (Galápagos rock-purslane) from extinction.

Professors and students from Ecuadorean universities have also been an important component of botanical research in the islands. It is common to see many of these scientists and their assistants working on projects in the terrestrial plants laboratory of the CDRS, or in the field. Indeed, botanical studies are now proceeding in the Galápagos Islands with a greater vitality than ever before.

The Galápagos Flora

According to Lawesson et al. (1987), the vascular flora of the Galápagos Islands comprises 863 members (species, subspecies, and varieties). Vascular plants possess conducting tissues known as *xylem* and *phloem*. These tissues transport water, dissolved minerals, and food throughout the plant's body. Vascular plants may be separated into two major categories: those that possess flowers (e.g., lilies, maples, roses, grasses), and those that do not (e.g., clubmosses, horsetails, ferns, conifers). Flowering plants, the topic of this book, are also known as *angiosperms*. This means "vessel seed" and refers to the fact that the seeds are borne inside a fruit. Some 749 members of the Galápagos flora fit this description.

Of the 749 members mentioned above, 216 (28.8%) are *endemic* to the archipelago. This means that they occur nowhere else in the world. These species, subspecies, and varieties typically resemble mainland plants, yet they are different. This makes sense when one considers that their ancestors originated on the mainland. After becoming established in the Galápagos Islands, they began to adapt to their new environment. Adaptation as a result of natural selection is the driving force behind *speciation*, the formation of new species. Over generations, offspring of these colonizers began to take on their own particular sets of characteristics. Several of these plants are now so different from their mainland counterparts that taxonomists consider them distinct species. Subspecies and varieties, on the other hand, are not sufficiently different to merit the rank of species. They are either more recent arrivals to the archipelago and have thus had less time to change (e.g., *Vallesia glabra* var. *pubescens*, pearl berry), or they have resulted from the subsequent evolution of an endemic species (e.g., the various subspecies of *Alternanthera filifolia*, thread-leafed chaff flower).

Seven Galápagos flowering plant groups have even been isolated long enough to constitute endemic genera. These are *Darwiniothamnus*, *Lecocarpus*, *Macraea*, and *Scalesia* of the family Asteraceae; *Brachycereus* and *Jasminocereus* of Cactaceae; and *Sicyocaulis* of Cucurbitaceae. In contrast, Hawaii boasts 32 endemic genera (Wagner et al. 1990), partly because Hawaii is older and more isolated from source floras.

A related idea is *adaptive radiation*, in which a particular animal or plant group adapts to a variety of habitats and, in the process, evolves into new entities. Darwin's finches are the classic example of this phenomenon. From a common ancestor, 13 species evolved by adapting to the food sources available in several unfilled ecological niches. The morphological evidence of these adaptations is best illustrated by their beaks. Certain species have thick beaks that can crush large seeds (Photo 11). Others have beaks better suited to feeding on leaves, flowers, or insects.

The genus *Scalesia* represents the plant kingdom's best example of adaptive radiation in the Galápagos. This endemic genus consists of 15 species, and a total of 19 members when subspecies and varieties are included. It has been suggested that the ancestor of these plants first colonized the arid lowlands of an island. This was followed by adaptation to moister sites and to other islands. In doing so, the leaves of these plants became markedly different from each other in shape and size. It seems that this is how they adapted to the humidity of their particular surroundings. Other Galápagos angiosperm genera demonstrating extensive adaptive radiation include *Chamaesyce* (nine endemic members, including eight species), *Mollugo* (nine endemic members, including four species), and *Opuntia* (14 endemic members, including six species).

Another 271 members (36.2%) of the angiosperm flora are *natives*. These plants colonized the Galápagos Islands by natural means (without human intervention) and are also found in other localities. A majority of the endemics and natives, perhaps 90%, probably originated in western South America. Other likely sources for a few members include the West Indies, Mexico, Central America, and North America (Porter 1983).

11. Darwin's finch feeding on a fruit of *Tournefortia rufo-sericea* (rufous-haired tournefortia)

Finally, Lawesson et al. (1987) mention 262 *exotic* flowering plants (35% of the angiosperm flora). These are present as a result of humans coming to the islands. Mauchamp (1997) has recently estimated that the actual number of exotics now stands at 438. The impact of these plants is discussed in a later section.

Arrival and Establishment

Long-Distance Dispersal

Being volcanic in origin, the Galápagos Islands were initially devoid of life, yet they are now populated by several hundred different kinds of plants. The explanation for this is twofold. The presence of exotics is the result of human contact. Some of these plants were introduced to provide food, while others were brought merely for aesthetic reasons. Still others arrived purely by accident, perhaps as a seed in the ballast of a ship.

Native plants and the ancestors of endemics, however, relied on a phenomenon known as *long-distance dispersal.* This refers to the natural transport of the plant itself, or its propagules (seeds or spores), from a point of origin to a new location some distance away. Dispersal to an oceanic island can be accomplished in a variety of ways. For example, many plants produce fruits or seeds that are eaten by birds (see Photos 62, 353). If such a bird happens to reach an island soon after ingesting the seeds, it might deposit them where they have a chance of survival. Studies have determined that many seeds are able to withstand passage through the digestive system of a bird without detrimental effects. In fact, this sometimes increases the germination ability of such seeds. Other plants have fruits or seeds that possess some form of adhesion, such as barbs or mucilage (see Photos 108, 245). These too may eventually reach the shore of a distant island while attached to the feathers of an exhausted bird. Still other seeds might arrive stuck to the muddy feet of a bird that frequents ponds or marshes.

Coastal plants typically arrive on the ocean currents. *Cocos nucifera* (coconut palm) and *Scaevola plumieri* (inkberry) are prime examples of plants with this type of dispersal mechanism (see Photos 111, 134). Most plants that take advantage of this method possess fruits that are not only buoyant but can also survive long periods in salt water. Occasionally, an entire plant might survive a trip to an island as part of a larger mass of floating vegetation that has been torn loose from a river bank on the continent. This type of dispersal, known as *rafting,* has also been used to explain how certain animals have reached the Galápagos.

Many plant propagules possess wings or plumes that allow them to be caught up in a passing breeze (Photo 12). Upon reaching sufficient altitude, they may travel hundreds of miles before falling back to earth. As chance has it, that bit of earth might be an oceanic island. Other seeds can become airborne simply because of their light weight. For example, orchid seeds are the size of dust particles and are carried throughout the world upon the winds.

Obviously, it has not been possible to observe the arrival of each plant or its propagule firsthand, so there is no exact record of the methods used. However, by studying the fruits and seeds of the resident species, and the distribution of each within the archipelago, one can ascertain the most likely means of dispersal. Porter (1983) suggested that birds have been responsible for 48.6% of the dispersal events, followed

12. Wind-dispersed fruits of *Porophyllum ruderale* var. *macrocephalum* (poreleaf)

by humans (38.8%), oceanic drift (7.0%), and wind (5.6%). If only natural agents are addressed, then the numbers are 79.8% for birds, 12.7% for ocean currents, and 7.5% for wind. Porter also stated that if the archipelago is 3 to 5 million years old, then one introduction every 7,300 to 12,100 years would account for the presence of the endemics and natives.

The flora of the Galápagos Islands is *disharmonic* in composition. This means that certain types of plants, known for their long-distance dispersal abilities, are typically present in greater numbers than those with poor dispersal techniques. Ferns (38 genera, 100 species) are found in great abundance in the archipelago because their spores are easily carried by the wind. Members of the families Amaranthaceae (7 genera, 29 species), Cyperaceae (7 genera, 35 species), and Poaceae (33 genera, 65 species) are highly represented because their seeds are a favorite source of food for many birds. In addition, they often grow in somewhat muddy areas. Any seeds that drop to the ground might stick to a bird's foot and be carried aloft. Another family with many representatives in the archipelago is Asteraceae (38 genera, 70 species). Its members often have fruits with plumelike hairs that allow them to be picked up by the wind. Others have fruits possessing barbs or mucilage that stick to the feathers of birds. This phenomenon of disharmony is often used to support the theory of the archipelago's oceanic origin. If the islands had been connected to the mainland in the past, we would expect there to be a more even representation of plant dispersal types. In other words, species with relatively poor dispersal mechanisms would be more common, since there would not have been an expanse of ocean to cross that necessitated long-distance dispersal capabilities.

Germination

Undoubtedly many seeds successfully arrived in the archipelago, only to be confronted by the obstacle of germination. It appears that the majority of plants that were successful in colonizing the arid lowlands have seeds that undergo a period of dormancy before germinating (McMullen 1986b). This seems only reasonable, as these seeds could wait out the cool, dry season and then germinate once the rains came. On the other hand, those plants inhabiting the moist uplands tend to germinate immediately.

Dormancy also seems to be correlated with dispersal mechanism. Seeds carried internally by birds show a higher percentage of dormancy than those carried externally. Perhaps this is because seeds passing through a bird's digestive system need a protective covering, which also acts as a dormancy mechanism.

Natives and endemics exhibit a greater degree of dormancy than exotics. While the latter appear better able to immediately exploit any habitat that opens up, natives and endemics display a "strategy" of waiting for optimum conditions before germinating. Many of the exotics produce a multitude of seeds, which could compensate for the high mortality that results from immediate seed germination in an inhospitable environment. This also explains why exotics are often successful in advancing into areas that are temporarily exposed due to such factors as fires and pasture use. These plants are preadapted to making a quick start in these disturbed locations and, in turn, slowing or halting the return of native vegetation.

Reproduction

All plants that were able to gain an initial foothold then faced the challenge of reproducing in order to become firmly established. Angiosperms are known to exhibit many different breeding strategies. However, the majority of those studied in the Galápagos Islands are *self-compatible* (McMullen 1987, 1990). This means that each individual possesses both male (stamen) and female (pistil) reproductive structures, either in the same flower or in separate flowers, and that the sperm produced by such an individual is capable of fertilizing its own eggs. What's more, most of these plants are *autogamous*. In other words, not only are they self-compatible, but their flowers actually pollinate themselves. The pollen grains may simply rub up against the receptive portion of the pistil (the stigma) as the flower opens. Or perhaps the reproductive structures are arranged so that gravity causes pollen to fall on the stigma. In any case, this method seems reasonable for an archipelago, since, hypothetically, only one individual would be necessary to start a sexually reproducing colony.

Based on the above discussion, it seems clear that most Galápagos angiosperms are able to reproduce without the aid of a pollinator. Yet there are insects available to these plants (Linsley et al. 1966; McMullen 1985, 1986a, 1989, 1994; McMullen and Naranjo 1994; McMullen and Viderman 1994). The most important of these is *Xylocopa darwini*, the endemic Galápagos carpenter bee. It is *polylectic*, which means that it visits many different plants for pollen and nectar. In fact, the flowers of 79 species of plants have been recorded as having been visited by this bee (Photo 13; see also 153).

Few other insects visit flowers in this archipelago (McMullen 1993). Those that have been observed doing so include butterflies, moths, ants, beetles, crickets, dipteran flies, roaches, hover flies, and stilt bugs (Photos 14–16). Not all of these insects are known to be effective in pollination. However, given that many of them perform this function on the mainland, it's not unreasonable to assume that they might also pollinate plants in the archipelago. Finches, doves, and mockingbirds also play a role in the pollination of

13. (left) Galápagos carpenter bee (*Xylocopa darwini*) visiting the flowers of *Prosopis juliflora* (mesquite)

14. (above) Galápagos blue butterfly (*Leptotes parrhasioides*) visiting the flowers of *Scalesia gordilloi* (Gordillo's scalesia)

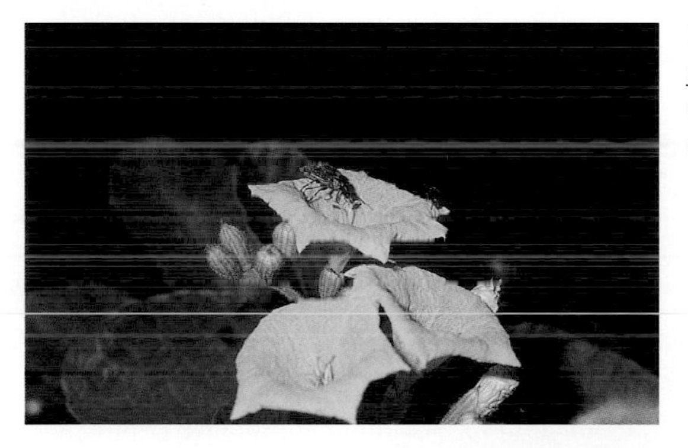

15. Beetles (*Perepitragus fuscipes*) visiting a flower of *Cordia lutea* (yellow cordia) at night

16. Fly (*Ornidia obesa*) visiting a flower of *Cordia leucophlyctis* (cordia)

17. Darwin's finch visiting the flowers of *Erythrina velutina* (flame tree)

some Galápagos angiosperms, especially on islands where the carpenter bee is absent (Grant and Grant 1981) (Photo 17).

Wind pollination is of little importance in the Galápagos (McMullen and Close 1993). This conclusion is based on pollen collection studies, the fact that most Galápagos flowers produce little pollen, the scarcity of pollen in core samples (Colinvaux and Schofield 1976), and the fact that most of the species flower during the time of year least suited for wind pollination (rainfall and garúa).

Based upon the results of breeding studies and the observation that there are relatively few pollinators in the Galápagos archipelago, it appears likely that the plants able to establish themselves early in the history of the islands were those that possessed upon arrival, or subsequently developed, the ability to self-pollinate. The initial absence of insect pollinators also explains the small flower size, drab appearance (whites and yellows), and limited pollen production of most of the endemic plants. Without faithful pollinators there would have been little or no selection for large showy flowers. In addition, the production of more pollen than necessary for autogamy would have been a waste of the limited resources available to the plant.

Orchids are especially interesting when discussing the establishment of plants on oceanic islands. Thousands of orchid species inhabit mainland Ecuador, and as mentioned earlier, their seeds are easily dispersed great distances by the wind. Yet only 14 species are found in the Galápagos. The explanation for this paradox may be twofold. First, orchids normally have a symbiotic relationship with a fungus. The name of this symbiosis, *mycorrhizae*, means "fungus-root" and refers to the close association between the orchid's roots and a fungus. In this relationship, the plant provides the fungus with carbohydrates produced by photosynthesis, while the fungus increases the plant's ability to absorb water and minerals from the soil. If an orchid's arrival in the archipelago is not preceded by the establishment of its fellow symbiont, it will either fail to germinate or will not reach maturity. Second, many orchid flowers have coevolved with a specialized pollinator, usually some type of insect. This means that the orchid typically cannot reproduce sexually unless its pollinator is present. Thus, it's possible that many orchid seeds reached the Galápagos over the eons, germinated, produced adult plants, and flowered, only to find that their required pollinator had not preceded them.

Vegetation Zones

The vegetation of the Galápagos Islands has traditionally been separated into six or seven altitudinal zones, and the southern slope of Santa Cruz in particular has often been used to demonstrate this pattern (Wiggins and Porter 1971; van der Werff 1979). Listed below is a brief description of each zone and representative plants that typically inhabit these areas.

Littoral (Coastal) Zone

This zone, which occupies the shoreline, varies in composition from lava boulders to sandy beaches (Photos 18–19). Common plants include *Avicennia germinans*

18. Littoral (Coastal) Zone on Santa Cruz: *Cryptocarpus pyriformis* (salt bush) and *Sesuvium portulacastrum* (common carpetweed)

19. Littoral (Coastal) Zone on Santa Cruz: *Avicennia germinans* (black mangrove), *Batis maritima* (saltwort), and *Sesuvium portulacastrum* (common carpetweed)

20. Salt crystals on a leaf of *Avicennia germinans* (black mangrove)

(black mangrove), *Batis maritima* (saltwort), *Conocarpus erectus* (button mangrove), *Cryptocarpus pyriformis* (salt bush), *Heliotropium curassavicum* var. *curassavicum* (seaside heliotrope), *Ipomoea pes-caprae* (beach morning-glory), *Laguncularia racemosa* (white mangrove), *Lycium minimum* (Galápagos lycium), *Maytenus octogona* (maytenus), *Nolana galapagensis* (Galápagos clubleaf), *Rhizophora mangle* (red mangrove), *Scaevola plumieri* (inkberry), *Sesuvium portulacastrum* (common carpetweed), *Sporobolus virginicus* (beach dropseed), and *Trianthema portulacastrum* (trianthema).

Plants that live in this zone must be capable of surviving in a salt-rich environment that may be covered with water during certain periods of the day but exposed at others. Thus, many have developed adaptations that allow them to persist. For example, *Avicennia germinans* (black mangrove) possesses many above-ground root extensions called *pneumatophores* that help the plant "breathe" by taking in oxygen. These structures are required due to the waterlogged condition of the soil. In addition, a layer of salt is often present on the upper surface of the leaves (Photo 20). It is excreted by specialized glands that provide the plant with a means of ridding itself of excess salt. As one might expect, many of the plants found here are adapted for water dispersal and are found on beaches throughout the world. Indeed, this ability to move from place to place by ocean currents explains why most of the plants inhabiting this zone are not endemics.

Arid Zone

This, the most extensive of all zones, is next to be encountered as one moves inland from the coast. As the name implies, it is extremely dry during most of the year, and many of the plants living here are *deciduous*. This means that the plants drop their leaves during the cool, dry season to reduce water loss. Because of this, the zone often appears to be occupied by dead or dying individuals. In reality, they are simply waiting until the rains of the warm season revive them. The most obvious members of this zone on many islands are *Opuntia echios* (prickly pear cactus) and *Jasminocereus thouarsii* (candelabra cactus) (Photo 21). Also prominent is *Bursera graveolens* (incense tree), a tree whose grayish bark gives a ghostly appearance to this zone (Photo 22). Trees that are less common but still characteristic include *Cordia lutea* (yellow cordia), *Croton scouleri* (Galápagos croton), *Erythrina velutina* (flame tree), and *Piscidia carthagenensis* (piscidia). On some islands the Arid Zone contains few or no trees. Instead, the entire landscape is dominated by plants such as *Brachycereus nesioticus* (lava cactus) and

21. Arid Zone on Santa Cruz: *Acacia macracantha* (acacia), *Cordia lutea* (yellow cordia), *Jasminocereus thouarsii* var. *delicatus* (candelabra cactus), and *Opuntia echios* var. *gigantea* (prickly pear cactus)

22. Arid Zone on Pinta: *Bursera graveolens* (incense tree)

Tiquilia nesiotica (gray matplant). Bartolomé is an example of such an island (Photo 23).

Shrubs are common as well, and walking through certain parts of this zone can be next to impossible due to their close spacing. Making this task even more imposing is the fact that many possess nasty spines and thorns that tear indiscriminately at clothing and flesh. Examples of common shrubs include *Acacia macracantha* (acacia), *Alternanthera echinocephala* (spiny-headed chaff flower), *Castela galapageia* (castela), *Chamaesyce viminea* (spurred chamaesyce), *Cordia revoluta* (revolute-leafed cordia), *Cryptocarpus pyriformis* (salt bush), *Gossypium darwinii* (Darwin's cotton), *Lantana peduncularis* (Galápagos lantana), *Maytenus octogona* (maytenus), *Parkinsonia aculeata* (Jerusalem thorn), *Prosopis juliflora* (mesquite), *Scutia spicata* var. *pauciflora* (thorn shrub), *Scalesia affinis* (radiate-headed scalesia), *Tournefortia psilostachya* (smooth-stemmed tournefortia), *T. pubescens* (white-haired tournefortia), and *Waltheria ovata* (waltheria).

Herbaceous representatives of the Arid Zone include *Aristida subspicata* (Galápagos three-awn grass), *Mentzelia aspera* (stickleaf), *Sarcostemma angustissimum* (Galápagos sarcostemma), and *Tribulus cistoides* (puncture weed). According to Wiggins and Porter (1971), this zone may extend inland to an elevation of over 120 meters on the southern slope of some islands, and up to 300 meters on the northern side.

23. Arid Zone on Bartolomé: *Tiquilia nesiotica* (gray matplant)

24. Transition Zone on Santa Cruz: *Opuntia echios* var. *gigantea* (prickly pear cactus), *Pisonia floribunda* (Galápagos pisonia), *Tournefortia rufo-sericea* (rufous-haired tournefortia), *Zanthoxylum fagara* (cat's claw)

Transition Zone

This zone (Photo 24) comprises both deciduous and evergreen trees such as *Bursera graveolens* (incense tree), *Cordia lutea* (yellow cordia), *Pisonia floribunda* (Galápagos pisonia), *Piscidia carthagenensis* (piscidia), and *Psidium galapageium* (Galápagos guava).

The shrub and herb layers are also well represented and include *Cardiospermum galapageium* (Galápagos heartseed), *Chiococca alba* (milkberry), *Clerodendrum molle* (glorybower), *Mormordica charantia* (bitter melon), *Plumbago scandens* (white leadwort), *Rhynchosia minima* (rhynchosia), *Senna occidentalis* (coffee senna), *Tournefortia pubescens* (white-haired tournefortia), and *T. rufo-sericea* (rufous-haired tournefortia).

Wiggins and Porter (1971) state that on the southern slope of Santa Cruz, this zone may reach approximately 200 meters altitude. On the northern and northeastern slopes of the larger islands it may occur even higher. This zone possesses many of the plants found in both the Arid and Scalesia Zones, but in fewer numbers. For example, *Opuntia echios* (prickly pear cactus) occupies the Transition Zone, but individuals are noticeably less common than in the Arid Zone. *Zanthoxylum fagara* (cat's claw) also inhabits this zone, but it becomes dominant only at higher elevations on some islands. For many, the Transition Zone is the toughest to pinpoint with certainty due to the lack of a few dominant indicator species.

Scalesia Zone

This zone, which extends to an elevation of 400–550 meters on the southern slope of Santa Cruz, is dominated by the evergreen tree *Scalesia pedunculata* (tree scalesia) (Wiggins and Porter 1971) (Photo 25). This same species forms the Scalesia Zone on Floreana, San Cristóbal, and Santiago. On Fernandina, however, this zone is dominated by *S. microcephala* var. *microcephala* (small-headed scalesia). Things are a bit more complex on Isabela. The Scalesia Zone on Volcán Alcedo and Volcán Darwin is composed of *S. microcephala* var. *microcephala*, while on Volcán Ecuador and Volcán Wolf it is composed of *S. microcephala* var. *cordifolia*. On Volcán Cerro Azul and Volcán Sierra Negra the zone is dominated by *S. cordata* (heart-leafed scalesia).

Additional trees common in the Scalesia Zone on Santa Cruz are *Pisonia floribunda* (Galápagos pisonia), *Psidium galapageium* (Galápagos guava), and *Zanthoxylum fagara* (cat's claw). During the cool season these forests are surrounded by the mist known locally as *garúa*. This provides moisture not only for the trees but also for numerous plants known as *epiphytes* that make their homes on the trunks and branches of these trees. Included among this group are *Epidendrum spicatum* (buttonhole orchid),

25. Scalesia Zone on Santa Cruz: *Scalesia pedunculata* (tree scalesia)

26. Scalesia Zone on Santa Cruz: *Scalesia pedunculata* (tree scalesia) and *Frullania* sp. (liverwort)

Ionopsis utricularioides (ionopsis), *Passiflora colinvauxii* (Colinvaux's passion flower), *P. suberosa* (passion flower), *Peperomia galapagensis* (Galápagos peperomia), *Phoradendron henslowii* (Galápagos mistletoe), and *Tillandsia insularis* (Galápagos tillandsia). Another group of epiphytes is equally noticeable but differs from the previously mentioned plants in that its members lack flowers. These are the various kinds of mosses, clubmosses, ferns, and liverworts. One of the more common of these is *Frullania* (liverwort). This brown-colored plant forms large clumps in the trees of this zone and is quite easy to identify (Photo 26).

Shrubs are well developed in this zone, but they are not as closely spaced as in the Transition Zone. Those that one is most likely to encounter include *Chiococca alba* (milkberry), *Darwiniothamnus lancifolius* (lance-leafed Darwin's shrub), *D. tenuifolius* (thin-leafed Darwin's shrub), *Psychotria rufipes* (white wild coffee), and *Tournefortia rufo-sericea* (rufous-haired tournefortia).

Herbs that best indicate the Scalesia Zone are *Adenostemma platyphyllum* (adenostemma), *Borreria laevis* (smooth borreria), *Diodia radula* (buttonweed), and *Justicia galapagana* (Galápagos justicia). Ferns belonging to the genera *Adiantum* (maidenhair fern) (Photo 27), *Asplenium* (spleenwort), *Doryopteris* (hand fern) (Photo 28), and *Polypodium* (polypody) are also common.

Zanthoxylum Zone

This zone is dominated by the small evergreen tree *Zanthoxylum fagara* (cat's claw) (Photo 29). However, shrubs and herbs are also abundant, and epiphytes crowd the tree

27. Scalesia Zone on Santa Cruz: *Adiantum* sp. (maidenhair fern)

28. Scalesia Zone on Santa Cruz: *Doryopteris pedata* var. *palmata* (hand fern)

29. Zanthoxylum Zone on Pinta: *Zanthoxylum fagara* (cat's claw)

limbs (Photo 30). In fact, this has been referred to as the "Brown Zone" due to the abundance of lichens that drape the trees and turn a brownish color during the cool season. Today, little of this zone persists, the majority having been eliminated after humans colonized the islands.

Miconia Zone

This zone, located on San Cristóbal and Santa Cruz, at one time consisted almost entirely of the shrub *Miconia robinsoniana* (Galápagos miconia) (Photo 31). Unfortunately, it is now greatly reduced due to burning, grazing by introduced animals, and competition from introduced plants. The worst among the latter is *Cinchona succirubra*

30. (right) Zanthoxylum Zone on Santa Cruz: *Tournefortia rufo-sericea* (rufous-haired tournefortia), *Zanthoxylum fagara* (cat's claw) draped with lichen *Ramalina usnea*

31. (below) Miconia Zone on Santa Cruz: *Miconia robinsoniana* (Galápagos miconia)

32. Miconia Zone on Santa Cruz: overgrown with *Cinchona succirubra* (quinine tree)

33. Miconia Zone on Santa Cruz: *Pteridium aquilinum* (bracken fern) and *Miconia robinsoniana* (Galápagos miconia)

(quinine tree) (Photo 32). Interspersed among the *Miconia* plants are several different types of clubmosses and ferns, the most notable being *Pteridium aquilinum* (bracken fern) (Photo 33). Among the flowering plants are *Cuphea racemosa* (white cuphea), *Hypericum uliginosum* var. *pratense* (St. John's wort), *Jaegeria gracilis* (Galápagos jaegeria), *Polygonum opelousanum* (knotweed), and *Verbena litoralis* (vervain). This zone begins at approximately 400–550 meters elevation (Wiggins and Porter 1971).

Fern-Sedge (Pampa) Zone

This zone, which consists primarly of clubmosses, ferns, sedges, and grasses (Photos 34–35), is found at the highest elevations on several of the larger islands. According to Wiggins and Porter (1971), the zone begins at an elevation of approximately 525–50 meters. No trees were located here before the arrival of humans. In fact, the tallest plant was *Cyathea weatherbyana* (Galápagos tree fern), which reaches a height of approximately 3 meters (Photo 36). During the cool season, this zone is almost continuously wet, and small pools of water are common. Some of these pools take on a mottled red and green appearance due to the presence of *Azolla microphylla* (water fern) on the water's surface (Photo 37). Other characteristic plants are *Ageratum conyzoides* (ageratum), *Cyperus grandifolius* (sedge), *C. virens* (sedge), *Habenaria monorrhiza*

34. (right) Fern-sedge Zone on Santa Cruz: *Lycopodium cernuum* (clubmoss) and *Jaegeria gracilis* (Galápagos jaegeria)

35. (below) Fern-sedge Zone on Santa Cruz

(fringed orchid), *Hypericum uliginosum* var. *pratense* (St. John's wort), *Jaegeria gracilis* (Galápagos jaegeria), *Lobelia xalapensis* (lobelia), *Ludwigia leptocarpa* (false loosestrife), and *Vigna luteola* (wild cowpea). As in the Miconia Zone, long-term survival of the native vegetation is threatened by the quinine tree.

Ecological Zones

Johnson and Raven (1973) suggest a different classification approach. This reduces the seven previously mentioned vegetation zones into three major ecological zones. These are the Littoral (Coastal) Zone; the Arid Lowlands, which include the Arid and

36. (left) Fern-sedge Zone on Santa Cruz: *Cyathea weatherbyana* (Galápagos tree fern)

37. (above) Fern-sedge Zone on Santa Cruz: *Azolla microphylla* (water fern) in temporary pond

Transition vegetation zones; and the Moist Uplands, which include the Scalesia, Zanthoxylum, Miconia, and Fern-Sedge vegetation zones. This system is more valuable for visitors to the Galápagos because the seven vegetation zones, though attractive in theory, are obvious only on a few of the higher islands. Indeed, they are not readily discernible on most of the islands visited by tourists. Therefore, a less technical approach that relates to relative location and moisture content seems advisable. These are the terms used in the "Habitat" category of the plant descriptions that follow.

Threats to the Vegetation

Introduced Animals

With the arrival of humans came the domesticated animals and plants upon which their survival often depends. Included among the former were pigs, goats, burros, and cattle (Photo 38). As is often the case, many of these began to roam about at will and eventually became wild. Such animals are referred to as *feral*, and their impact is twofold. First, their movements trample and wear trails through the vegetation. A few individuals might not be reason for concern, but in large numbers they can cause tremendous damage. Second, each of these animals feeds primarily on plants. The effect of this browsing on endemic and native Galápagos plants has been well documented.

Enclosures have been erected on a few islands to protect threatened plants from permanent damage. For example, early in 1997, two quadrats in the highlands of Volcán Alcedo (Isabela) were enclosed by fences. One of these was 500 square meters, the other 300 square meters. This was done primarily to protect *Cyathea weatherbyana* (Galápagos tree fern). In addition, many separate individuals of *Tournefortia rufo-sericea* (rufous-haired tournefortia) and *Zanthoxylum fagara* (cat's claw) were enclosed.

38. Cattle on Santa Cruz

These three species are especially important in providing shade for the giant tortoises that inhabit Volcán Alcedo. In addition, they tend to collect moisture on their leaves, which later drips to the ground and forms shallow pools that are used by the tortoises. It is anticipated that the enclosed individuals will survive and, moreover, that they will provide a source of spores and seeds that may be used to reestablish these plants in areas that have been cleared of feral animals. Although somewhat effective, these methods do not address the real problem. Ultimately, more permanent solutions must be implemented, as the following examples demonstrate.

In the late 1950s, fishermen released a small number of goats on Pinta. Their expectation was that the goats might serve as a source of fresh meat on future visits. However, the outcome of this "experiment" proved tragic for the island's vegetation. These goats thrived and reproduced at an astonishing rate, and their effect on the plant life was devastating. Extensive areas were severely damaged, and the entire island was in danger until a hunting campaign was initiated. It's estimated that as many as 40,000 goats were killed during the following years. Fortunately, disaster was averted and the plants are making a comeback. My assistants and I spent the summers of 1990 and 1993 on Pinta without observing a single living goat. Vigilance must be kept, however, as a few goats have been sighted on this island since then.

Unfortunately, all of the resources of the CDRS and GNPS cannot be used to solve the problems of a single island. Isabela is currently facing a similar crisis with goats and burros, the population of the former having been estimated at over 100,000. Once again, the result has been catastrophic. Areas that were once covered with lush highland vegetation are now dry, parched earth. As goat herds pass by, they create clouds of dust. This destruction of plant life also affects the giant tortoises that share this island. As mentioned earlier, the water dripping off the leaves of these plants during the garúa season produces the tortoises' wading ponds. Fortunately, Project Isabela was organized to rid the island of its feral animals. In 1996 alone, approximately 25,300 goats and 680 burros were destroyed on Volcán Alcedo. If this effort is to be successful, these animals must never again be allowed to reach such numbers.

A more recent threat to Galápagos plants involves an insect known as the cottony cushion scale (*Icerya purchasi*) (Photo 39). Since its discovery, it has been observed on several native plants, including *Cordia lutea* (yellow cordia), *Hibiscus tiliaceus* (seaside hibiscus), *Laguncularia racemosa* (white mangrove), *Merremia aegyptica* (hairy merremia), *Parkinsonia aculeata* (Jerusalem thorn), and *Piscidia carthagenensis* (piscidia). Infested plants tend to become sickly and frequently die. As is often the case with exotic organisms, this insect appears to have no natural enemies in the archipelago. An answer might be found in California. The cottony cushion scale was introduced into this state, probably from Australia, in the 1800s and threatened to destroy its citrus industry. However, a ladybird beetle (*Rodolia cardinalis*) and cryptochetum fly (*Cryptochetum iceryae*) were found to control the spread of the scale. This is an example of *biological control*, a methodology that has never been used in the Galápagos. Obviously, studies must be conducted to determine if indeed there are no native enemies of the scale living in the archipelago. Next, scientists must make sure that the ladybird beetle and cryptochetum fly will not feed detrimentally on native insects. Further steps will not be taken until it is certain that these introduced control agents will not become pests themselves. Certainly there are risks, but there is also hope that the destruction of Galápagos plants by the cottony cushion scale can be halted, or at least limited.

39. Cottony cushion scale (*Icerya purchasi*) on Santa Cruz

Exotic Plants

The invasion by exotic plants, now estimated at 438 species (Mauchamp 1997), is perhaps more insidious. Approximately 26% of these are *introduced weeds*, which means that they were brought to the islands unintentionally. For example, seeds might arrive hidden among a ship's ballast. If the ship foundered, or if part of its ballast were discarded, then the seeds might have an opportunity to germinate. Other seeds might be carried in fodder for livestock, on the fur or feathers of domesticated animals, or on the clothing or hair of their owners. It's even possible that seeds might make the journey inside the gut of one of these animals. If the seeds were voided after reaching the islands, they might germinate and start a population.

Many other plants were brought to the islands specifically to be cultivated for food. Examples include *Ananas comosus* (pineapple), *Carica papaya* (papaya), *Citrus limetta* (sweet lime), *Passiflora quadrangularis* (giant granadilla), *Persea americana* (avocado), and *Psidium guajava* (common guava). Others were carried to the islands by early settlers to be used as ornamentals. Included among these are *Allamanda cathartica* (golden trumpet), *Bougainvillea spectabilis* (bougainvillea), *Catharanthus roseus* (Madagascar periwinkle), *Delonix regia* (flamboyant), *Hibiscus rosa-sinensis* (Chinese hibiscus), *Lantana camara* (multicolored lantana), and *Nerium oleander* (common oleander). Residents had more specific uses in mind for still other plants. *Cedrela odorata* (Spanish cedar) and *Ochroma pyramidale* (balsa) were planted for their wood; *Pennisetum purpureum* (elephant grass) was brought to the islands as a forage plant; and *Cinchona succirubra* (quinine tree) was originally cultivated to produce quinine, a well-known treatment for malaria. Cultivated species account for approximately 74% of all exotics in the islands (Mauchamp 1997).

Approximately half (55.3%) of the 438 exotics appear to be of little concern at this time (Mauchamp 1997). In other words, they are typically found only where originally planted, and they are not competing with the endemic and native members of the Galápagos flora. A second group, 42.2%, has become *naturalized*. Included among these are those introduced weeds, crop plants, and ornamentals that have moved out among the surrounding environs and become established. The majority of these are not causing problems at this time. However, 2.5% of the exotics have become *invasive* and are aggressively outcompeting Galápagos endemics and natives. Considered among the worst are *Cinchona succirubra* (quinine tree), *Kalanchoe pinnata* (air plant), *Lantana camara* (multicolored lantana), *Pennisetum purpureum* (elephant grass), *Psidium guajava* (common guava), and *Rubus niveus* (hill raspberry). These plants are capable of causing tremendous damage, especially on the inhabited islands.

Between 1987 and 1995, the number of exotic plant species on Santa Cruz jumped from 197 to 348. This represents an increase of 76.6%. During the same time period, there was a 28.1% increase on Floreana (89 to 114), a 60.9% increase on San Cristóbal (110 to 177), and a 64.7% increase on Volcán Sierra Negra (Isabela) (68 to 112) (Mauchamp 1997). Not only do certain exotics compete directly with endemics and natives for space and resources, but they might also be responsible for the introduction of other pest organisms. It's not at all unreasonable to suspect that the cottony cushion scale arrived in the archipelago on a crop plant or ornamental.

The number of residents, tourists, and scientists is sure to increase in the years ahead. This in turn increases the chances of exotics reaching the shores of the uninhabited islands. Seeds are easily carried on the boot soles or in the gear of these individuals. Warnings by the CDRS and CNPS to avoid such transfers may not be sufficient. Ultimately, limiting the number of visitors and permanent residents or establishing an effective quarantine program will be required if this onslaught of exotics is to be controlled. This will continue to be a major challenge for the CDRS and GNPS; both organizations are currently involved in efforts to reverse this trend and save the archipelago's unique vegetation.

Reasons for Hope

Occasionally there are success stories, such as that involving *Scalesia atractyloides* (scalesia). Restricted to Santiago, this endemic was thought by many to have vanished forever, primarily as a result of grazing by goats. However, in 1995, GNPS wardens reported finding five individuals of this species. Another example relates to *Linum cratericola* (Floreana flax). This endemic member of the Linaceae was discovered in 1966 by Uno Eliasson. It was known from only two locations and was last seen in 1981. Given the negative effects of feral goats and burros, and the abundance of *Lantana camara* (multicolored lantana) on the island, most botanists thought *L. cratericola* to be extinct after repeated searches. However, in April 1997, Alan Tye, the resident botanist at the CDRS, rediscovered a small population (13 individuals) of this species at one of the original locations. Obviously, these species remain endangered, but the CDRS is now attempting to ensure that they do not disappear forever. The moral boost given the CDRS and GNPS personnel upon making such discoveries cannot be overstated.

Plant Descriptions

Plant Key

Trees with Alternate Leaves and White Flowers

Scientific Name: Scalesia pedunculata Hook. f. (Photos 40–41)
Common Names: Tree scalesia, *lechoso*

Family: Asteraceae (Sunflower)

Range: Endemic

Islands Inhabited: Floreana, San Cristóbal, Santa Cruz, Santiago

Habitat: Moist uplands

Description: Tree to ca. 20 m tall. *Leaves* alternate or almost opposite, simple; blade ovate-lanceolate to ovate, ca. 6–20 cm long, both surfaces hairy and glandular, margins entire or occasionally toothed. *Flowers* in discoid heads, disc 1–2 cm or more across, flowers white, ca. 50–150 or more per head, perfect. *Fruit* an achene, flattened, 4–6 mm long, rarely with a short awn on top, to ca. 1 mm long; seed 1.

Comments: Scalesia is one of seven angiosperm genera endemic to the Galápagos. It is composed of shrubs and trees that possess soft wood and gummy sap, and the young branches are typically somewhat hairy. The leaves are usually crowded near the branch tips, and dried remnants of the previous year's growth often persist beneath. The inflorescences, known as "heads," are borne singly or in small clusters (2–10) near the ends of the branches. The genus is particularly notable in being one of the few members of the family Asteraceae that includes trees. Most are either shrubs or herbs, and common examples include dandelions, sunflowers, and daisies.

Forests of *S. pedunculata* are well represented near Los Gemelos and the Tortoise Reserve (Santa Cruz), and these areas are where the species is most easily observed. Additional stands, located on Santa Cruz's northern slope, may be viewed from the top of El Puntudo (see Photo 25). These forests dominate what is known as the "Scalesia Zone." This tree is also well represented on Floreana. However, on San Cristóbal and Santiago it is not nearly as common as in the past. This situation appears to be the result of damage caused by goats and pigs that were allowed to roam at will; it demonstrates the danger of introducing alien species to the archipelago.

Scalesia atractyloides Arn. is an endangered species known only from Santiago. It is a small tree that reaches a maximum height of 3 m. Its leaves are alternate or almost opposite, and its leaf blades are linear or lanceolate in shape and typically 6–10 cm long. The leaf margins are entire. Each discoid head is 0.7–1.5 cm across and contains ca. 40–100 white flowers. This species consists of two varieties. *Scalesia atractyloides* var. *atractyloides* has linear to linear-lanceolate leaf blades with short hairs on the upper surface, whereas var. *darwinii* (Hook. f.) Eliasson has lanceolate to linear-lanceolate leaf blades with long white hairs on both surfaces.

40. (above) *Scalesia pedunculata* (tree scalesia), flowers

41. (right) *Scalesia pedunculata* (tree scalesia), flowers

Scientific Name: Ceiba pentandra (L.) Gaertn. (Photo 42)
Common Names: Silk-cotton tree, *ceibo*

Family: Bombacaceae (Bombax)

Range: Cultivated escape; also known from other tropical regions throughout the world.

Islands Inhabited: Floreana, Isabela (SN), Santa Cruz

Habitat: Moist uplands

Description: Tree to 40 m tall, larger trunks often with vertical supports known as "buttresses" at the base, trunk and branches often with irregular-shaped spines, those of the trunk 1–3 cm long. *Leaves* alternate, palmately compound; leaflets typically 5–7, oblanceolate to oblong-obovate, 8–20 cm long, margins usually minutely serrate, occasionally entire. *Flowers* terminal, solitary; corolla white to pink, petals 5, 1.5–2 cm long, covered with white to yellow hairs on the outside; stamens 5. *Fruit* a capsule, elliptic, 10–12 cm long; seeds numerous, brown, embedded in a large mass of silky, white to tan-colored hairs.

Comments: Several impressive examples of this tree may be seen near Santo Tomás (Isabela). In many parts of the world it is grown for its silky, cottonlike hairs known commercially as "kapok." This material has been used for insulation and as a filler in softballs, pillows, and life preservers.

42. *Ceiba pentandra* (silk-cotton tree)

Scientific Name: Ochroma pyramidale (Cav. ex Lam.) Urb. (Photo 43)
Common Name: Balsa

Family: Bombacaceae (Bombax)

Range: Cultivated escape; also known from other regions of tropical America.

Islands Inhabited: Isabela (SN), Santa Cruz

Habitat: Moist uplands

Description: Tree to ca. 30 m tall, trunk somewhat buttressed at the base; bark smooth, gray with white spots or streaks. *Leaves* alternate, simple; blade slightly 3-lobed, broadly ovate, 15–30 cm long, margins entire. *Flowers* axillary, solitary; corolla white, 10–15 cm long, petals 5; stamens numerous, united into a central column. *Fruit* a capsule, linear-oblong, 15–23 cm long; seeds numerous, surrounded by dense brownish fibers.

Comments: According to Wiggins and Porter (1971), this tree was introduced into the Galápagos in 1940. It was cultivated for its light wood known as "balsa," but that product has never become a commercial success in the archipelago. This tree is easy to identify, not only by the characters mentioned above but also by the fact that its leaves are usually confined to the top part of the tree, forming a flattened crown. This is quite obvious from a distance when traveling through the highlands.

43. *Ochroma pyramidale* (balsa)

Scientific Name: Bursera graveolens (Kunth) Triana & Planch.

(Photos 44–46)

Common Names: Incense tree, *palo santo*

Family: Burseraceae (Torchwood)

Range: Native; also known from western South America (Venezuela to Peru).

Islands Inhabited: Darwin, Española, Fernandina, Floreana, Genovesa, Isabela (A,CA,D, E,SN,W), Marchena, Pinta, Rábida, San Cristóbal, Santa Cruz, Santa Fe, Santiago, Islet(s)

Habitat: Arid lowlands

Description: Tree or shrub 3–12 m tall, branches reddish brown, upper branches spreading. *Leaves* alternate but usually clustered at the branch tips, odd-pinnately compound, 10–28 cm long; main stalk winged; leaflets usually 5–7, ovate to ovate-lanceolate, bright green and more or less smooth, margins serrate to crenate, lateral leaflets to 6 cm long, terminal leaflet slightly longer. *Flowers* unisexual (plants dioecious), in axillary panicles 10–13 cm long; corolla pale yellowish white, 4 mm long, petals 4; staminate flowers with 4 stamens; pistillate flowers losing petals before staminate flowers. *Fruit* a drupe, brown when mature, roundish, 9–12 mm long; seed 1.

Comments: The specific epithet, *graveolens*, means "strong-smelling." This refers to the plant's aromatic resin, which produces a strong but pleasant odor when a branch is broken. The common name *palo santo* means "holy stick" and comes from the fact that the branches are often burned for incense in churches. On certain islands, such as Pinta, this tree is often heavily draped with the lichen *Ramalina usnea* (Photo 44). During the cool season, both *Bursera graveolens* and *B. malacophylla* lose their leaves and appear lifeless, taking on a whitish gray or almost silver color. However, they are simply dormant and will become active when the rains return.

44. (top) *Bursera graveolens* (incense tree), draped with the lichen *Ramalina usnea*

45. (middle) *Bursera graveolens* (incense tree), flowers

46. (bottom) *Bursera graveolens* (incense tree), flowers

Scientific Name: Bursera malacophylla B. L. Rob. (Photo 47)
Common Names: Galápagos incense tree, Galápagos *palo santo*

Family: Burseraceae (Torchwood)

Range: Endemic

Islands Inhabited: Baltra, Daphne, Santa Cruz, Seymour

Habitat: Arid lowlands

Description: Small tree or shrub to 4 m tall, branches reddish brown, young branch tips with woolly hairs. *Leaves* alternate but usually clustered at the branch tips, odd-pinnately compound, 8–15 cm long; main stalk winged; leaflets 5–9, typically ovate to somewhat roundish, grayish green and covered with woolly hairs, margins roughly crenate, lateral leaflets to 6 cm long, terminal leaflet slightly longer. *Flowers* unisexual (plants dioecious), in axillary panicles 9–15 cm long; corolla pale yellowish white, 4 mm long, petals 4; staminate flowers with 4 stamens; pistillate flowers losing petals before staminate flowers. *Fruit* a drupe, brown or black when mature, roundish, 8–13 mm long; seed 1.

Comments: Bursera malacophylla is considered rare. However, it can be observed on Seymour, where it often grows in close proximity to *Opuntia echios* var. *zacana* (prickly pear cactus). The flowers of both *B. malacophylla* and *B. graveolens* (incense tree) turn orange as they become dry.

47. *Bursera malacophylla* (Galápagos incense tree), *Opuntia echios* var. *zacana* (prickly pear cactus) in the foreground

Scientific Name: Carica papaya L. (Photos 48–49)
Common Name: Papaya

Family: Caricaceae (Papaya)

Range: Cultivated escape; also known from other tropical regions throughout the world, originally from tropical America.

Islands Inhabited: Floreana, Isabela (SN), San Cristóbal, Santa Cruz

Habitat: Arid lowlands

Description: Small tree or shrub 2–8 m tall, trunk well marked with leaf scars, containing a milky sap. *Leaves* alternate, clustered near the top of the tree and forming a distinctive crown, simple, to 60 cm wide, palmately 5- to 7-lobed, each lobe secondarily pinnately lobed. *Flowers* unisexual (plants dioecious or occasionally monoecious). Staminate flowers in axillary cymes to ca. 1 m long; corolla yellowish white, tubular with 5 lobes, tube 1.5–2 cm long, lobes 1–1.5 cm long; stamens 10. Pistillate flowers in axillary clusters of 1–3; corolla yellowish white, somewhat campanulate, petals 5, 5–7 cm long. *Fruit* a berry, yellowish orange at maturity, variously shaped, to 30 cm or more in length; seeds numerous, black.

Comments: This tree is best known for its fleshy fruits, which are located on the trunk just below the leaves and are eaten fresh or made into juice. However, if too ripe, these fruits begin to have an odor and taste that some may find offensive. Papayas have another important use. They contain papain, a digestive enzyme that is used commercially as a meat tenderizer. In addition, the fruits have been suggested for treating the sting of a jellyfish: Rubbing the fruit's flesh on a sting is said to ease the pain. This tree can be found near any of the settled areas in the archipelago, and the fruits can be purchased at the local markets. After dark, its flowers are often visited by the carmine hawk moth (*Agrius cingulatus*).

48. (above) *Carica papaya* (papaya)

49. (right) *Carica papaya* (papaya), flowers

Scientific Name: Terminalia catappa L. (Photo 50)
Common Names: Tropical almond, *almendra*

Family: Combretaceae (Combretum)

Range: Cultivated; also known from other regions of tropical America, originally from Malaysia.

Islands Inhabited: Isabela (SN), Santa Cruz

Habitat: Arid lowlands

Description: Tree to 25 m tall, branches arranged in a regular series of whorls. *Leaves* alternate but usually clustered near the branch tips, simple, broadly obovate, 10–30 cm long, dark glossy green, margins entire. *Flowers* perfect or staminate (plants polygamous), in terminal spikes 10–25 cm long. Perfect flowers at the base of each spike; calyx greenish white, tubular with campanulate top, 5-lobed, 4–5 mm across; corolla absent; stamens 10. Staminate flowers at the top of each spike; calyx greenish white, tubular with campanulate top, 5-lobed, 2 mm across; corolla absent; stamens 10. *Fruit* drupelike, greenish gray, sometimes with a reddish tinge, ellipsoid, 2–6 cm long, slightly winged, middle layer somewhat fibrous; seed 1.

Comments: In many parts of tropical America this tree is planted along streets and in yards for the shade it provides. The leaves usually turn bright red before falling to the ground. Its seeds are edible and provide an oil that is important in some countries. The genus name is derived from the Latin *terminus*, referring to the leaves clustered near the branch tips.

50. *Terminalia catappa* (tropical almond), flowers and fruits

Scientific Name: Croton scouleri Hook. f. (Photos 51–52)
Common Names: Galápagos croton, *chala*

Family: Euphorbiaceae (Spurge)

Range: Endemic

Islands Inhabited: Daphne, Darwin, Española, Fernandina, Floreana, Genovesa, Isabela (A,CA,D,E,SN), Marchena, Pinta, Rábida, San Cristóbal, Santa Cruz, Santa Fe, Santiago, Seymour, Wolf, Islet(s)

Habitat: Arid lowlands and moist uplands

Description: Small tree or shrub 2–6 m tall. *Leaves* alternate, sometimes crowded near the branch tips, simple; blade linear to ovate, 2–25 cm long, both surfaces with hairs, usually stellate, margins entire or with minute teeth. *Flowers* unisexual (plants dioecious), in terminal racemes. Staminate corolla yellowish white, petals 5, 1.8–3.6 mm long; stamens 14–22. Pistillate flowers with 5 sepals, greenish, 1.4–2 mm long, densely hairy; corolla usually absent. *Fruit* a capsule, roundish, slightly 3-lobed, 5–7 mm in diameter; seeds 3, yellowish, pitted.

Comments: Croton scouleri consists of four varieties (Wiggins and Porter 1971), which are extremely difficult to distinguish from one another. Fortunately, their distributions within the archipelago help with this task. For example, var. *brevifolius* (Andersson) Müll. Arg. occurs only on Floreana. It is considered rare and is typically found at higher elevations.

Both var. *darwinii* G. L. Webster and var. *grandifolius* Müll. Arg. are also considered rare. The former inhabits Darwin, Santa Cruz, and Wolf and appears to prefer lower elevations. The latter is found on Darwin, Isabela (SN), Pinta, Rábida, San Cristóbal, Santa Cruz, and Santiago, and it typically occupies higher elevations.

Croton scouleri var. *scouleri* (Photos 51–52), the most common member of the species, occurs on Daphne, Española, Fernandina, Floreana, Genovesa, Isabela (A, CA,D,E,SN), Marchena, Pinta, Rábida, San Cristóbal, Santa Cruz, Santa Fe, Santiago, Seymour, and one or more of the smaller islands. It is typically found at lower elevations. Several finch species rely on the seeds of this plant as a food source.

Representatives on Santa Cruz provide the greatest challenge for successful identification, as three of the four varieties occur here. However, knowledge of a few additional facts should help with their identification. For example, var. *darwinii* has leaf blades that are elliptic to roundish and typically 3–7 cm long and 1.5–6 cm wide. In addition, its seeds are 3.9–4.5 mm long. *Croton scouleri* var. *grandifolius* has ovate-shaped leaf blades that are 6–15 cm long and 3–8 cm wide and seeds that are 3.7–4.7 mm long. *Croton scouleri* var. *scouleri* possesses leaf blades that vary from linear to ovate and are 2–12 cm long and 0.8–2.3 cm wide. Its seeds are usually 2.6–3.6 mm long.

Croton scouleri contains a resin that produces a dark stain on any cloth that it touches. No doubt, all park wardens and most scientists are quite familiar with the results of accidentally rubbing their clothes and backpacks against these plants. The stains are impossible to remove.

Members of the genus *Croton* should not be confused with the common garden croton, *Codiaeum variegatum* (L.) Juss. (Photo 53). Although members of the same family, Euphorbiaceae, they are quite different. Garden croton has leaves that are usually much larger, somewhat leathery, and marked with white, yellow, or red. This shrub is common in gardens and yards on Santa Cruz. However, all of its parts are toxic and should not be tasted.

51. (top) *Croton scouleri* var. *scouleri* (Galápagos croton), flowers
52. (bottom left) *Croton scouleri* var. *scouleri* (Galápagos croton), flowers
53. (bottom right) *Codiaeum variegatum* (garden croton), flowers

Scientific Name: Piscidia carthagenensis Jacq. (Photos 54–56)
Common Names: Piscidia, *matazarno*

Family: Fabaceae (Pea)

Range: Native; also known from other regions of tropical America (Mexico and the West Indies to the northwestern countries of South America).

Islands Inhabited: San Cristóbal, Santa Cruz

Habitat: Arid lowlands and moist uplands

Description: Tree or shrub to ca. 15 m tall. *Leaves* alternate, odd-pinnately compound; leaflets 7–13, obovate, elliptic, or ovate, ca. 4–20 cm long. *Flowers* in axillary or lateral panicles; corolla white to pink or purplish, 1.3–1.8 cm long, composed of 1 large standard petal, 2 lateral wing petals, and 2 lower keel petals that are somewhat fused; stamens 10. *Fruit* a legume, linear and flattened with 4 longitudinal wings, 5–11 cm long, wings 1.3–1.8 cm wide; seeds 1–8, reddish brown.

Comments: The extremely hard inner wood of *Piscidia carthagenensis* is used for constructing houses and boats in the archipelago. Interestingly, another member of this genus, found in the West Indies, is used as a source of fish poison.

54. (left) *Piscidia carthagenensis* (piscidia)

55. (top right) *Piscidia carthagenensis* (piscidia), flowers

56. (bottom right) *Piscidia carthagenensis* (piscidia), fruits

Scientific Name: Inga schimpffii Harms (Photo 57)
Common Names: Silky inga, *guaba de machete*

Family: Mimosaceae (Mimosa)

Range: Cultivated escape; also known from western South America

Islands Inhabited: Isabela (SN), Santa Cruz

Habitat: Moist uplands

Description: Tree to ca. 10 m tall, young branches with a few fine, silky hairs. *Leaves* alternate, even-pinnately compound; main axis winged, 11–25 cm long; leaflets 6–8, oblong to obovate-oblong, 7–25 cm long, somewhat leathery, with small glands present at the base. *Flowers* in axillary clusters resembling panicles, white; corolla tubular with 5 lobes, ca. 1.5 cm long; stamens numerous, ca. 3 cm long, united at the base to form a tube. *Fruit* a legume, elongate, 30–50 cm long, 4–5.5 cm wide, somewhat compressed; seeds numerous, covered with a whitish pulp.

Comments: Inga edulis Mart. (woolly inga), also cultivated, differs from *I. schimpffii* in having young branches that are densely covered with woolly hairs. In addition, its leaves are not so leathery in texture, and its fruits are only ca. 1 cm wide. It is known from Floreana and Santa Cruz. According to Wiggins and Porter (1971), both of these species have been used to provide shade for cacao trees on mainland plantations. In addition, the whitish pulp surrounding the seeds is edible and is thought by some to be quite tasty.

57. *Inga schimpffii* (silky inga), flowers and fruit

Scientific Name: Citrus limetta Risso (Photo 58)
Common Names: Sweet lime, *lima dulce*

Family: Rutaceae (Citrus)

Range: Cultivated escape; also known from other tropical regions throughout the world, originally probably from India and Asia.

Islands Inhabited: Floreana, San Cristóbal, Santa Cruz

Habitat: Moist uplands

Description: Small tree to 8 m tall, branches glandular-dotted, with axillary, brown-tipped thorns 1.5–7.5 cm long. *Leaves* alternate, compound but reduced to 1 leaflet, obovate, ovate, or elliptic, 5.5–17 cm long, somewhat leathery, margins minutely crenate; petiole winged, spatulate, 2–8 mm wide. *Flowers* in axillary clusters; corolla white, 2–5 cm across, petals 5, ca. 1.6 cm long; stamens numerous. *Fruit* a hesperidium, yellow when mature, ovoid, ca. 5 cm in diameter, containing a juicy, greenish pulp; seeds numerous.

Comments: According to Wiggins and Porter (1971), *Citrus limetta* is the only member of the genus that has become naturalized in the archipelago. The fruit's taste has been described by some as lacking zest, but it is quite popular in Latin America. Other members of this genus that are cultivated in the Galápagos Islands include *C. aurantifolia* (Christm.) Swingle (lime, *lima*), *C. aurantium* L. (sour orange, *naranja*), *C. limon* (L.) Burm. f. (lemon, *limón*), *C. medica* L. (citron, *citrón*), *C. paradisi* Macfad. (grapefruit, *toronja*), *C. reticulata* Blanco (mandarin orange, *mandarina*), and *C. sinensis* (L.) Osbeck (sweet orange, *naranja*) (Photo 59). Each of these is found on Santa Cruz. In addition, grapefruits and sweet oranges are grown on San Cristóbal.

58. (top) *Citrus limetta* (sweet lime), fruit

59. (bottom) *Citrus sinensis* (sweet orange), fruits

Scientific Name: Sapindus saponaria L. (Photo 60)
Common Names: Soapberry, *jaboncillo*

Family: Sapindaceae (Soapberry)

Range: Native; also known from other tropical regions throughout the world, originally from tropical America.

Islands Inhabited: Isabela (SN), Santa Cruz, Santiago

Habitat: Moist uplands

Description: Tree to ca. 20 m tall, branches with tiny brown lenticels. *Leaves* alternate, even- or odd-pinnately compound, 12–30 cm long; petiole and main axis somewhat winged; leaflets 8–11, obovate to elliptic, 4–14 cm long, margins entire. *Flowers* unisexual (plants monoecious), in terminal compound panicles; flower 4–5 mm across, corolla white, petals 4–5; staminate flowers with 8 stamens. *Fruit* a schizocarp of 1–2 drupelike sections, obovoid, 1–2.5 cm in diameter; seeds 1 per section.

Comments: This plant's genus name comes from the Latin *sapo*, "soap," and *indicus*, "Indian," which refers to the fruit pulp, used at one time as a substitute for soap. Wagner et al. (1990) mention that the seeds of this species have been used in Hawaii to make necklaces.

60. *Sapindus saponaria* (soapberry), flowers and fruits

See also

Acnistus ellipticus (p. 118)
Brugmansia candida (p. 119)
Cordia alliodora (p. 61)
Leucaena leucocephala (p. 66)
Melia azedarach (p. 76)
Scalesia affinis (p. 103)
S. aspera (p. 103)
S. baurii (p. 104)
S. cordata (p. 106)
S. divisa (p. 105)
S. helleri (p. 150)
S. microcephala (p. 106)
S. stewartii (p. 108)
S. villosa (p. 109)
Solanum erianthum (p. 121)
Vallesia glabra (p. 99)

Trees with Alternate Leaves and Yellow or Orange Flowers

Scientific Name: Cascabela thevetia (L.) Lippold (Photo 61)
Common Name: Yellow oleander

Family: Apocynaceae (Dogbane)

Range: Cultivated; also known from other tropical regions throughout the world, originally from tropical America.

Islands Inhabited: Santa Cruz

Habitat: Arid lowlands

Description: Small tree to 7 m tall, branches containing a milky sap. *Leaves* alternate, simple, linear-lanceolate, 10–15 cm long, margins entire. *Flowers* in terminal cymes; corolla yellow, funnelform with 5 lobes, to 5 cm across; stamens 5. *Fruit* drupelike; seeds 2–4.

Comments: Although uncommon, yellow oleander may be observed in the town of Puerto Ayora. The genus name appears to have come from the Spanish *cascabela*, meaning "small bell," and refers to the shape of the flower. A cardiac glucoside (heart depressant) is obtained from this species (Wagner et al. 1990).

61. *Cascabela thevetia* (yellow oleander), flowers

Scientific Name: Cordia lutea Lam. (Photos 62–63; See also Photo 15)
Common Names: Yellow cordia, *muyuyo*

Family: Boraginaceae (Borage)

Range: Native; also known from mainland Ecuador and Peru.

Islands Inhabited: Baltra, Española, Floreana, Genovesa, Isabela (A,CA,D,SN), Marchena, Pinta, Pinzón, Rábida, San Cristóbal, Santa Cruz, Santa Fe, Santiago, Seymour, Wolf, Islet(s)

Habitat: Arid lowlands

Description: Small tree or shrub to 8 m tall, young branches covered with hairs. *Leaves* alternate, simple; blade ovate to somewhat roundish, 4–10 cm long, upper surface rough, lower surface hairy, margins minutely crenate. *Flowers* in cymes; corolla yellow, funnelform with 5–8 lobes, 2–4 cm across; stamens 5–8. *Fruit* a drupe, white, roundish, 8–12 mm across; seeds 1–4.

Comments: This is one of seven Galápagos species in the genus *Cordia*, which was named for the German botanists Euricius (1485–1535) Cordus and his son Valerius (1515–44). Of these, *C. lutea* is the easiest to identify, due to its showy yellow flowers. Not only are they beautiful, but they also possess a wonderfully sweet aroma. In addition, they are one of the favorite sources of nectar for the Galápagos sulfur butterfly (*Phoebis sennae*) and the Galápagos carpenter bee (*Xylocopa darwini*). The carpenter bee's method of removing nectar, called "nectar robbing," involves making a small slit at the base of the corolla. Then, without ever entering the flower, it sucks out the nectar. This behavior is referred to as robbing because the bee receives its reward, but no pollination occurs. Sometimes the pollen itself is used as a food source. Both beetles (*Amblycerus piurae*) and spiders (*Anyphaenoides octodentata*) have been observed feeding on it during the night. The fruits of this plant are well known for their fleshy pulp, which, when crushed, is at first slimy but soon becomes extremely sticky. On a practical note, this juicy material can be used to seal envelopes. The fruit is pleasingly sweet when first tasted, but soon becomes bitter. This may be a purely human assessment, as Galápagos mockingbirds (*Nesomimus parvulus*) appear quite fond of the fruits. Rats have also been observed feeding on the fruits, sometimes for hours at a time during the night. Wood from this tree is used for carvings that are sold locally at souvenir shops.

Cordia alliodora (Ruiz & Pav.) Oken (laurel), known only from Santa Cruz, is an introduced tree that can reach more than 20 m in height. Its leaves are more or less elliptic in shape and are larger than those of any other Galápagos cordia. The young stems, leaves, and inflorescences typically possess stellate hairs. The calyx is 5-lobed but distinctly 10-ribbed. The white corolla is 5-lobed, ca. 1.2 cm across, and has 5 stamens. The fruit is cylindrical in shape and is dispersed within the dried corolla.

62. (above) *Cordia lutea* (yellow cordia), flowers and fruits

63. (right) *Cordia lutea* (yellow cordia), flowers

Scientific Name: Parkinsonia aculeata L. (Photos 64–65)
Common Names: Jerusalem thorn, *palo verde*

Family: Caesalpiniaceae (Caesalpinia)

Range: Native; also known from other tropical regions throughout the world, originally from tropical America.

Islands Inhabited: Baltra, Española, Floreana, Isabela (SN), Pinzón, San Cristóbal, Santa Cruz, Seymour, Islet(s)

Habitat: Arid lowlands

Description: Small tree or shrub 2–10 m tall, with smooth green bark, branches with spines that are actually modified leaf parts. *Leaves* alternate, bipinnately compound with 2–4 ribbonlike pinnae, 17–45 cm long; leaflets 20–30 pairs per pinna, oblong to elliptic, 3–10 mm long. *Flowers* in axillary racemes 6–20 cm long; corolla yellow with one petal splotched reddish orange, 2–3 cm across, petals 5; stamens 10. *Fruit* a legume, linear, 5–20 cm long, strongly constricted between the seeds; seeds 1–6, shiny, brownish.

Comments: This beautiful tree, with its delicate drooping branches, is one of the first plants noticed by arriving passengers as they leave the airport. The flowers are a favorite nectar source of the Galápagos carpenter bee (*Xylocopa darwini*). In addition, finches often feed on the flowers and leaflets. When feeding on the latter, they simply pull a leaf up to the branch upon which they are perched, hold it with their foot, and bite off the leaflets one by one. Jerusalem thorn may be familiar to visitors from the southern United States, where it has become naturalized. The genus is named for the British botanist John Parkinson (1567–1650), author of the last major herbal in the seventeenth century.

64. (above) *Parkinsonia aculeata* (Jerusalem thorn), flowers

65. (left) *Parkinsonia aculeata* (Jerusalem thorn), flowers

Scientific Name: Persea americana L. (Photo 66)
Common Names: Avocado, *aguacate*

Family: Lauraceae (Laurel)

Range: Cultivated escape; also known from other tropical regions throughout the world, originally from Mexico and Central America.

Islands Inhabited: Floreana, Isabela (SN), San Cristóbal, Santa Cruz, Santiago

Habitat: Moist uplands

Description: Tree to ca. 30 m tall, branches numerous. *Leaves* alternate, usually crowded near the branch tips, simple; blade ovate-oblong, 10–30 cm long, leathery, upper surface dark green and shiny, lower surface pale green and dull, margins entire. *Flowers* in axillary and terminal panicles; tepals 6, yellowish, hairy, 3–6 mm long; fertile stamens 9. *Fruit* a drupe, yellowish green to purplish, ovoid or pear-shaped, 7–20 cm long, fleshy with a leathery rind; seed 1.

Comments: This tree is cultivated for its nutritious fruits, which have a thick yellow flesh that takes on a buttery consistency when mature. These fruits are eaten alone, as side dishes, and in salads. Unfortunately, pigs and cattle also find the fruits irresistible and have introduced this plant into some areas of the national park. This is troublesome, not only because the avocados may crowd out native and endemic trees and shrubs but also because their large leaves can prevent the growth of herbaceous species beneath.

66. *Persea americana* (avocado), fruits

Scientific Name: Cedrela odorata L. (Photos 67–68)
Common Names: Spanish cedar, *cedro cubano*

Family: Meliaceae (Mahogany)

Range: Cultivated escape; also known from other regions of tropical America (West Indies and South America), originally from the West Indies.

Islands Inhabited: Floreana, San Cristóbal, Santa Cruz

Habitat: Moist uplands

Description: Tree to ca. 30 m tall, bark smooth. *Leaves* alternate, even-pinnately compound; leaflets 10–20, ovate-lanceolate, to ca. 12 cm long, margins entire. *Flowers* in panicles; corolla yellowish, petals 4 or 5; stamens 4–6. *Fruit* a woody capsule, to ca. 4 cm long, splitting into 5 sections when mature; seeds numerous, winged below.

Comments: Spanish cedar is an extremely valuable timber tree and is harvested for its aromatic, reddish brown wood, which is used in shipbuilding. Although humans find the odor pleasant, it repels insects. Thus, the wood has been used for closets, furniture, and boxes of various types, including cigar boxes. This tree has escaped cultivation in some areas and may become a problem if not controlled.

67. (above) *Cedrela odorata* (Spanish cedar)

68. (right) *Cedrela odorata* (Spanish cedar), fruits

Scientific Name: Acacia insulae-iacobi Riley (Photo 69)
Common Names: Acacia, *algarrobo*

Family: Mimosaceae (Mimosa)

Range: Native; also known from Ecuador, Bolivia, and Argentina.

Islands Inhabited: Baltra, San Cristóbal, Santa Cruz, Santiago

Habitat: Arid lowlands

Description: Small tree to ca. 7 m tall, branches with straight stipular spines to ca. 2 cm long. *Leaves* alternate, bipinnately compound with 1–5 pairs of pinnae; leaflets 5–10 pairs per pinna, oblong, 3–15 mm long and 1–4 mm wide. *Flowers* in axillary, round clusters ca. 1 cm across, orange-yellow; corolla tubular with 5 lobes, ca. 3 mm long; stamens numerous. *Fruit* a legume, linear, 10–17 cm long, slightly flattened; seeds 3–15.

Comments: This tree is commonly found growing on lava and in rough, rocky areas. *Leucaena leucocephala* (Lam.) de Wit (lead tree) (Photo 70) and members of the genus *Mimosa* also have leaves that are bipinnately compound and flowers that are arranged in round clusters. *Leucaena leucocephala* is an introduced shrub or small tree that occurs on San Cristóbal. It differs from *Acacia* in having white flowers and no spines. Three members of the genus *Mimosa* also occur on San Cristóbal, as cultivated escapes. They differ from *Acacia* and *Leucaena* in having white to pinkish flowers and spines on the branches.

69. (left) *Acacia insulae-iacobi* (acacia), flowers and fruits

70. (right) *Leucaena leucocephala* (lead tree), flowers and fruits

Scientific Name: Acacia nilotica (L.) Delile (Photo 71)
Common Names: Nile acacia, *algarrobo*

Family: Mimosaceae (Mimosa)

Range: Cultivated escape; also known from other tropical regions throughout the world, originally from northwestern Africa.

Islands Inhabited: Santa Cruz

Habitat: Arid lowlands

Description: Small tree to ca. 8 m tall, branches with straight stipular spines to ca. 6 cm long. *Leaves* alternate, bipinnately compound with 3–16 pairs of pinnae; leaflets 10–30 pairs per pinna, oblong, 3–5 mm long and 1 mm wide. *Flowers* in axillary, round clusters ca. 6–15 mm across, yellow; corolla tubular with 5 lobes, 3–4 mm long; stamens numerous. *Fruit* a legume, linear, 10–25 cm long, flattened, often with constrictions between the seeds; seeds numerous.

Comments: The obviously constricted fruits of *Acacia nilotica* are quite different from those of other members of the genus. This makes for easy identification. When no fruits are present, however, observe the stalk holding one of the flower clusters. Near the middle of this stalk will be a few tiny bracts. These are not present at this location on the other species. In some parts of the world, especially India, the tree is used for timber. Although uncommon, it may be seen by visitors to Puerto Ayora (Santa Cruz). The genus name comes from the Greek *akis*, "sharp point," referring to the stipular spines characteristic of these plants.

71. *Acacia nilotica* (Nile acacia), flowers

Scientific Name: Acacia rorudiana Christoph. (Photo 72)
Common Names: Galápagos acacia, *algarrobo*

Family: Mimosaceae (Mimosa)

Range: Endemic

Islands Inhabited: Española, Floreana, Isabela (A,D,SN), Santa Cruz, Santiago

Habitat: Arid lowlands

Description: Small tree or shrub to ca. 8 m tall, branches somewhat twisting, with straight stipular spines to ca. 4 cm long. *Leaves* alternate, bipinnately compound with 5–30 pairs of pinnae; leaflets 14–24 pairs per pinna, oblong, 0.5–1 mm long and 0.2–0.4 mm wide, margins with fine hairs. *Flowers* in axillary, round clusters ca. 6–7 mm across, yellow to orange; corolla tubular with 5 lobes, 1–2 mm long; stamens numerous. *Fruit* a legume, linear-oblong, 7–16 cm long, flattened; seeds numerous.

Comments: Acacia macracantha Willd. closely resembles *A. rorudiana* but has slightly larger flower clusters (8 mm across) and larger leaflets (1–3 mm long and 0.5–1 mm wide). It occurs on Floreana, Isabela (D), Pinzón, San Cristóbal, Santa Cruz, Santiago, and one or more of the smaller islands. Some taxonomists believe that this native and *A. rorudiana* should actually be treated as one variable species. However, this decision awaits further studies.

72. *Acacia rorudiana* (Galápagos acacia), flowers

Scientific Name: Prosopis juliflora (Sw.) DC. (Photos 73–74; see also Photo 13)
Common Names: Mesquite, *algarrobo*

Family: Mimosaceae (Mimosa)

Range: Native; also known from other tropical regions throughout the world, originally from tropical America.

Islands Inhabited: Baltra, Española, Floreana, Isabela (CA,SN), Pinta, Pinzón, Rábida, San Cristóbal, Santa Cruz, Santa Fe, Santiago, Islet(s)

Habitat: Coastal zone and arid lowlands

Description: Small tree or shrub to ca. 10 m tall, branches with straight stipular spines to ca. 1.5 cm long. *Leaves* alternate, bipinnately compound with 1–2 (occasionally 3) pairs of pinnae; leaflets 6–20 pairs per pinna, oblong, 5–24 mm long and 1.5–6 mm wide, somewhat leathery. *Flowers* in axillary spikes ca. 5–12 cm long, greenish yellow; petals 5, 3–4 mm long; stamens 10. *Fruit* a legume, linear, straight or curved, 9–25 cm long, often with a flexible point at the tip, occasionally somewhat constricted between the seeds; seeds numerous.

Comments: Without flowers or fruits, *Prosopis juliflora* may be mistaken for *Acacia insulae-iacobi.* However, the young twigs are distinctly different. Those of *A. insulae-iacobi* possess numerous spots known as "lenticels" on the bark, while twigs of *P. juliflora* do not. Mesquite flowers are frequently visited by the Galápagos carpenter bee (*Xylocopa darwini*).

73. (above) *Prosopis juliflora* (mesquite), flowers

74. (left) *Prosopis juliflora* (mesquite), flowers and immature fruits

Scientific Name: Zanthoxylum fagara (L.) Sarg. (Photo 75; see also Photos 29–30)
Common Names: Cat's claw, uña de gato

Family: Rutaceae (Citrus)

Range: Native; also known from other regions of tropical America (Mexico and Florida to northwestern South America and the West Indies).

Islands Inhabited: Daphne, Española, Fernandina, Floreana, Isabela (A,CA,D,SN), Pinta, Rábida, San Cristóbal, Santa Cruz, Santiago, Islet(s)

Habitat: Arid lowlands and moist uplands

Description: Small tree or shrub to 10 m tall, branches with paired, hooked spines 2–6 mm long. *Leaves* alternate, odd-pinnately compound, 3.5–11 cm long; petiole winged, ca. 2 mm wide; main stalk winged and grooved on top, 2–4.5 mm wide; leaflets 5–11, obovate to broadly elliptic, ca. 1–3 cm long, somewhat leathery, margins minutely crenate. *Flowers* unisexual (plants dioecious) in axillary spikes ca. 5 mm long; corolla typically greenish yellow, petals 4, ca. 1 mm long; staminate flowers with 4 stamens. *Fruit* a follicle, dark brownish green, obovoid and beaked, 3.5–5 mm in diameter, somewhat leathery; seed 1, shiny black, round, 4 mm in diameter.

Comments: The genus derives its name from the Greek *xanthos*, "yellow," and *xylon*, "wood," alluding to the yellowish wood produced by many of its members. This tree is a common element of the highlands on many of the islands, and it often forms impenetrable thickets. Personal experience can attest to the appropriateness of this plant's common name. Its spines, which are as sharp as any cat's claws, can make hiking a torturous experience.

75. *Zanthoxylum fagara* (cat's claw), fruits

Scientific Name: Trema micrantha (L.) Blume (Photos 76–77)
Common Names: Trema, *niguito*

Family: Ulmaceae (Elm)

Range: Native; also known from other regions of tropical America.

Islands Inhabited: Isabela (A,CA,D,SN), Pinta, Santa Cruz, Santiago

Habitat: Arid lowlands

Description: Small tree or shrub to 10 m tall, branches spreading and somewhat drooping, young branches hairy. *Leaves* alternate, simple; blade lanceolate to lance-ovate, to 15 cm long, drooping, both surfaces somewhat rough, margins minutely serrate. *Flowers* unisexual (plants monoecious), in axillary cymes. Calyx of staminate flowers greenish yellow, sepals 5, 1.2–1.4 mm long; corolla absent, stamens 5. Calyx of pistillate flowers greenish yellow, 3 mm across, sepals 5; corolla absent. *Fruit* a drupe, roundish, 3–3.5 mm long, flesh cream-colored to reddish, juicy; seed 1, pale brown, with longitudinal ridges.

Comments: The wood of *Trema micrantha* does not appear to be of any commercial value in the archipelago.

76. (above) *Trema micrantha* (trema)

77. (left) *Trema micrantha* (trema), flowers

See also

Erythrina poeppigiana (p. 74)
E. velutina (p. 74)

Hibiscus tiliaceus (p. 128)
Tamarindus indica (p. 73)

Trees With Alternate Leaves and Pink, Red, or Purple Flowers

Scientific Name: Bauhinia monandra Kurz (Photo 78)
Common Name: Butterfly flower

Family: Caesalpiniaceae (Caesalpinia)

Range: Cultivated; also known from other regions of tropical America.

Islands Inhabited: Santa Cruz

Habitat: Arid lowlands

Description: Tree or shrub to 12 m tall. *Leaves* alternate, simple, deeply 2-lobed, broadly ovate, 8–20 cm long. *Flowers* in terminal racemes; corolla pinkish with one petal splotched purplish red, 10–12 cm across, petals 5; fertile stamen 1. *Fruit* a legume, oblong, 17–23 cm long; seeds numerous, shiny black.

Comments: Both *Bauhinia monandra* and *B. variegata* L. (poor man's orchid, *orquídea de pobre*) are used as ornamentals. The latter, also known only from Santa Cruz, differs in having racemes that are attached to the branch tips laterally rather than terminally. In addition, each flower has 5 fertile stamens, and the corolla is usually more of a purplish color, with 1 petal splotched dark purple. Interestingly, another form of *B. variegata*, perhaps the most popular of all, has completely white flowers. The genus name honors John and Caspar Bauhin, sixteenth century Swiss botanists. The 2 lobes of each leaf are said to represent these scientists.

78. *Bauhinia monandra* (butterfly flower), flowers

Scientific Name: Delonix regia (Bojer ex Hook.) Raf. (Photos 79–80)
Common Names: Flamboyant, royal poinciana, *flamboyán*

Family: Caesalpiniaceae (Caesalpinia)

Range: Cultivated escape; also known from other tropical regions throughout the world, originally from Madagascar.

Islands Inhabited: Santa Cruz

Habitat: Arid lowlands and moist uplands

Description: Tree to ca. 10 m tall, upper branches forming a spreading crown. *Leaves* alternate, bipinnately compound with 10–20 pairs of pinnae; leaflets 20–40 pairs per pinna, oblong, 5–10 mm long. *Flowers* in axillary racemes; corolla red with one petal splotched white and yellow, 8–15 cm across, petals 5; stamens 10. *Fruit* a legume, dark brown, flattened and oblong, 20–70 cm long, 3–7 cm wide, becoming somewhat woody; seeds 18–45, brown.

Comments: Visitors to the archipelago during this tree's flowering period are in for a treat. The blossoms are truly striking and are noticeable from quite a distance. However, even without flowers, it is recognizable due to the large seed pods. The species has been planted extensively around Puerto Ayora, especially in the town square.

Another cultivated escape belonging to this family is *Tamarindus indica* L. (tamarind, *tamarindo*) (Photo 81). This tree has pinnately compound leaves, each with 10–18 leaflets. Its pale yellow flowers are approximately 2–2.5 cm across, and its fruit pods are up to 15 cm long. This species is important in some countries as an ornamental and as a source of fruit for juices and preserves. It is known from Floreana, Isabela (SN), San Cristóbal, and Santa Cruz.

79. (above) *Delonix regia* (flamboyant), flowers

80. (left) *Delonix regia* (flamboyant), fruits

81. *Tamarindus indica* (tamarind)

Scientific Name: Erythrina velutina Willd. (Photos 82–83; see also Photo 17)
Common Names: Flame tree, caco

Family: Fabaceae (Pea)

Range: Native; also known from other regions of tropical America (West Indies to northern South America).

Islands Inhabited: Darwin, Genovesa, Isabela (W), Santa Cruz, Santiago, Wolf

Habitat: Arid lowlands

Description: Tree to ca. 12 m tall, with spines to 1.5 cm long, young stems covered with hairs. *Leaves* alternate, odd-pinnately compound; leaflets 3, somewhat triangular, 4–16 cm long, 4–19 cm wide, lower surface covered with hairs. *Flowers* in axillary racemes; corolla orange-red, flower 4–6 cm long, composed of 1 large standard petal (3–3.5 cm long and usually bent backward), 2 lateral wing petals, and 2 lower keel petals that are somewhat fused; stamens 10, fused into a tube. *Fruit* a legume, somewhat twisted, 7.5–14 cm long, constricted between the seeds; seeds usually 1–4, red with a short black line.

Comments: The genus derives its name from the Greek *erythros*, "red." This is in reference to the beautiful flowers, which normally appear during the cool season, when the leaves are absent. They are one of the favorite sources of nectar and pollen for the Galápagos carpenter bee (*Xylocopa darwini*) and several finch species. In fact, the activity at these trees during the flowering period, by both bees and finches, can only be described as intense. In addition, carpenter bees are known to build their nests in the tree's soft wood.

Other species found in the archipelago include *Erythrina corallodendron* L. (Santa Cruz) (Photo 84); *E. edulis* Triana (Volcán Sierra Negra of Isabela); *E. fusca* Lour. (Santa Cruz), commonly called *palo prieto*; *E. poeppigiana* (Walp.) O. F. Cook (Floreana, Isabela [SN], San Cristóbal, Santa Cruz), known as *poro gigante*; and *E. smithiana* Krukoff (Santa Cruz), known as *porotillo*. These are cultivated trees and are often planted in coffee and cacao plantations on the mainland to provide shade. In the Galápagos they are frequently used by farmers as "living fence posts." Branches cut from adult trees are put into the ground, and wire is strung between them. Before long, these fence posts develop new roots and branches. Leaves from these trees are used to feed Galápagos tortoises at the CDRS. Each of these species produces red flowers, except *E. poeppigiana*, whose flowers are orange.

82. (top) *Erythrina velutina* (flame tree), flowers

83. (middle) *Erythrina velutina* (flame tree), flowers

84. (bottom) *Erythrina corallodendron* (erythrina), flowers

Scientific Name: Melia azedarach L. (Photos 85–86)
Common Names: Chinaberry, *jazmin*

Family: Meliaceae (Mahogany)

Range: Cultivated escape; also known from other tropical regions throughout the world, originally from southwestern Asia.

Islands Inhabited: Floreana, San Cristóbal, Santa Cruz

Habitat: Arid lowlands

Description: Tree to ca. 10 m tall. *Leaves* alternate, bipinnately compound, 18–23 cm long; leaflets numerous, lanceolate to elliptic or ovate, 3–6.5 cm long, margins serrate. *Flowers* in axillary panicles; corolla typically pale purple, occasionally whitish, petals usually 5, occasionally 6, 8–10 mm long; stamens 10–12, the filaments all fused together to form a dark purple cylindrical tube. *Fruit* a drupe, yellow, roundish, 1.5–2 cm in diameter; seeds 1–6.

Comments: Although cultivated as an ornamental, this species often inhabits disturbed areas such as roadside ditches. Visitors may find it in and around Puerto Ayora (Santa Cruz), where its attractive fragrance often gives its presence away. Occasionally an individual tree will have completely white flowers, including the fused filaments. Wagner et al. (1990) mention that in Hawaii, this species was used to treat lepers' sores. Chinaberry is an aggressive weed in some parts of the tropics, so its presence in the archipelago should be monitored.

85. *Melia azedarach* (chinaberry), flowers and fruits

86. *Melia azedarach* (chinaberry), flowers

See also

Ceiba pentandra (p. 46)
Hibiscus rosa-sinensis (p. 135)
H. schizopetalus (p. 136)
Piscidia carthagenensis (p. 55)

Trees with Alternate Leaves and Green Flowers

Scientific Name: Conocarpus erectus L. (Photos 87–88)
Common Names: Button mangrove, *mangle boton, jelí*

Family: Combretaceae (Combretum)

Range: Native; also known from other tropical regions throughout the world (Mexico and Florida to Ecuador and Brazil, western Africa).

Islands Inhabited: Isabela (A,CA,SN), San Cristóbal, Santa Cruz, Santiago

Habitat: Coastal zone

Description: Small tree or shrub 3–8 m tall. *Leaves* alternate, simple, oblanceolate to narrowly elliptic, 2.5–9 cm long, somewhat leathery, with a pair of raised glands at the base, small pits where the secondary veins meet the major vein on the lower surface, margins entire or somewhat wavy. *Flowers* in spherical heads borne in terminal panicles or racemes; calyx greenish, cup-shaped with 5 lobes, 2 mm long; corolla absent; stamens usually 5, sometimes 10, much longer than the calyx. *Fruit* drupelike, reddish brown, 3–3.5 mm long, flattened and winged, tightly clustered into somewhat spherical "cones," 8–16 mm across; seed 1.

Comments: In order to observe the largest individuals of *Conocarpus erectus*, one needs to visit the town of Puerto Villamil (Isabela). These plants are of such prominence in this town that a street has been named for the genus. Its name comes from the Greek *konos*, "cone," and *carpos*, "fruit." This species is referred to as *C. erecta* L. in Wiggins and Porter (1971).

87. (above) *Conocarpus erectus* (button mangrove), flowers

88. (left) *Conocarpus erectus* (button mangrove), fruits

Scientific Name: Hippomane mancinella L. (Photo 89)
Common Names: Poison apple, *manzanillo*

Family: Euphorbiaceae (Spurge)

Range: Native; also known from the West Indies and surrounding mainland coasts (Mexico and Florida to Colombia).

Islands Inhabited: Floreana, Isabela (CA,D,SN,W), San Cristóbal, Santa Cruz, Santiago

Habitat: Coastal zone, arid lowlands, and moist uplands

Description: Tree to 10 m tall, containing milky sap. *Leaves* alternate, simple, ovate to elliptic, 3–7 cm long, margins entire. *Flowers* unisexual (plants monoecious), in terminal spikes 3–10 cm long; each group of flowers with a reddish brown glandular bract beneath. Staminate flowers in numerous clusters; calyx greenish, ca. 1 mm long, 2-lobed; corolla absent; stamens 2. Pistillate flowers 1–2 at base of spike; calyx greenish, ca. 3 mm long, 3-lobed; corolla absent. *Fruit* drupelike, yellowish green, ca. 3 cm in diameter, containing milky sap; seeds 6–9.

Comments: As with most members of the spurge family, this plant's milky latex can cause severe skin irritation upon contact. In addition, it can lead to temporary blindness. Stewart (1911) went so far as to warn his readers not to stand underneath this tree during a rain, as water from the leaves could cause great pain if it accidentally dripped into one's eyes. Although the fruit resembles an apple, it is extremely poisonous and should never be tasted. Despite these deadly attributes, this plant has become a popular ornamental around settled areas. Poison apple is normally an inhabitant of the coastal zone and arid lowlands, but it is occasionally found in the moist uplands.

89. *Hippomane mancinella* (poison apple), flowers and fruit

See also

Croton scouleri (p. 53)
Maytenus octogona (p. 139)
Ricinus communis (p. 141)
Urera caracasana (p. 142)

Trees with Opposite Leaves and White Flowers

Scientific Name: Avicennia germinans (L.) L. (Photos 90–91; see also Photos 19–20)
Common Names: Black mangrove, *mangle negro*

Family: Avicenniaceae (Black mangrove)

Range: Native; also known from other regions of tropical America (Texas and Florida to Peru and Brazil).

Islands Inhabited: Baltra, Española, Fernandina, Floreana, Isabela (D,SN), Pinzón, Rábida, San Cristóbal, Santa Cruz, Santiago, Islet(s)

Habitat: Coastal zone

Description: Tree or shrub to 25 m tall, branches spreading. *Leaves* opposite, simple; blade lanceolate to elliptic, 4.5–15 cm long, often with a grayish fuzz on the lower surface, margins entire. *Flowers* in terminal or axillary clusters; corolla white or pale yellow, with a yellow throat, campanulate with 4 lobes, to 1 cm across; stamens 4. *Fruit* a capsule, yellow at first but turning purplish black, oblong, 1.5–5 cm long; seed 1.

Comments: Black mangrove fruits are dispersed by ocean currents, and for this reason the species is found throughout the archipelago. Typically these plants possess numerous above-ground root extensions called "pneumatophores" that help the plant "breathe" by taking in oxygen. These structures, which may be up to 10 cm in length, are necessary due to the waterlogged condition of the soil in which this plant grows. A thin layer of salt is often observed on the upper surface of the leaves. It is excreted by specialized glands that are an adaptation to the extreme salinity of the mangrove's environment. The flowers, which have a pleasant odor, appear to be favorites of the recently introduced wasp *Polistes versicolor.* These insects constantly visit the flowers of black mangrove at Tortuga Bay (Santa Cruz).

90. (above) *Avicennia germinans* (black mangrove)

91. (left) *Avicennia germinans* (black mangrove), flowers

Scientific Name: Laguncularia racemosa (L.) Gaertn. f. (Photos 92–93)
Common Names: White mangrove, *mangle blanco*

Family: Combretaceae (Combretum)

Range: Native; also known from other tropical regions throughout the world (Mexico and Florida to Peru and Brazil, western Africa).

Islands Inhabited: Fernandina, Floreana, Isabela (A,CA,SN,W), Pinta, Pinzón, Rábida, San Cristóbal, Santa Cruz, Santiago, Islet(s)

Habitat: Coastal zone

Description: Small tree or shrub 2–10 m tall. *Leaves* opposite, simple, oblong to oval, ca. 5.5–9 cm long, somewhat leathery, with a pair of raised glands at the base, numerous tiny pits near the margins on both surfaces, margins entire or slightly wavy. *Flowers* usually perfect (occasionally staminate, then plants functionally dioecious), in axillary or terminal paniculate spikes 7–8 cm long; corolla white, ca. 2 mm long, petals 5; stamens 10. *Fruit* drupelike, grayish green, approximately 1.5–2 cm long, somewhat flattened and winged, with longitudinal ribs, topped by a persistent calyx, 2 mm long; seed 1.

Comments: The middle layer of this plant's fruit is filled with a "spongy" material that makes it an ideal candidate for oceanic dispersal. For this reason, white mangroves are common to tropical beaches in many parts of the world. As with *Rhizophora mangle* (red mangrove), the seeds often sprout while still on the parent plant, a condition known as "vivipary." However, the young seedlings will eventually drop off and begin a life on their own. White mangroves often produce numerous aboveground root extensions called "pneumatophores" that help them "breathe." *Avicennia germinans* (black mangrove) also demonstrates this phenomenon.

92. (above) *Laguncularia racemosa* (white mangrove), flowers

93. (right) *Laguncularia racemosa* (white mangrove), fruits

Scientific Name: Psidium galapageium Hook. f. (Photo 94)
Common Names: Galápagos guava, *guayabillo*

Family: Myrtaceae (Myrtle)

Range: Endemic

Islands Inhabited: Fernandina, Isabela (A,CA,SN), Pinta, San Cristóbal, Santa Cruz, Santiago

Habitat: Arid lowlands and moist uplands

Description: Small tree or shrub to 8 m tall. *Leaves* opposite, simple; blade elliptic to ovate, 1.8–5.5 cm long, margins entire. *Flowers* axillary, solitary, ca. 1–1.5 cm across; corolla white, petals 5, 4–9 mm long; stamens numerous, filaments white. *Fruit* a berry, yellow at first, turning reddish brown to black, roundish, 6–13 mm in diameter; seeds numerous.

Comments: This is one of two species of *Psidium* found in the archipelago, and it includes two varieties. *Psidium galapageium* var. *howellii* D. M. Porter (Photo 94), considered rare, occurs on San Cristóbal and Santa Cruz. It possesses flowers that are ca. 1 cm across, petals that are 4–5.5 mm long, and buds that have 5 lobes at their tips.

Psidium galapageium var. *galapageium* is found on Fernandina, Isabela, Pinta, Santa Cruz, and Santiago. It differs from var. *howellii* in having slightly larger flowers (to 1.5 cm across), longer petals (8–9 mm long), and flower buds that are not lobed.

94. *Psidium galapageium* var. *howellii* (Galápagos guava), fruits

Scientific Name: Psidium guajava L. (Photos 95–96)
Common Names: Common guava, *guayaba, guayabo*

Family: Myrtaceae (Myrtle)

Range: Cultivated escape; also known from other tropical regions throughout the world, originally from tropical America.

Islands Inhabited: Floreana, Isabela (CA,SN), San Cristóbal, Santa Cruz

Habitat: Moist uplands

Description: Tree to 8 m tall, young branches distinctly 4-angled. *Leaves* opposite, simple; blade elliptic to oblong-elliptic, 5–14 cm long, somewhat leathery, veins on upper surface impressed, margins entire. *Flowers* axillary, solitary or in clusters of 2–3, ca. 2.5 cm across; corolla white, petals 5, ca. 1–2 cm long; stamens numerous, filaments white. *Fruit* a berry, pale yellow, roundish to pear-shaped, to ca. 5 cm in diameter; seeds numerous.

Comments: The fruits of this tree contain a pink to cream-colored pulp that locals use to make preserves and juice. These fruits are also enjoyed by domesticated animals such as pigs and cattle, as well as by many birds. This is how the seeds are spread throughout each island. Common guava is one of the major threats to the native vegetation on the inhabited islands, especially on Isabela and San Cristóbal, where it has formed dense stands by outcompeting the native and endemic plant species.

95. (above) *Psidium guajava* (common guava), flowers

96. (left) *Psidium guajava* (common guava), fruit

Scientific Name: Syzygium jambos (L.) Alston (Photo 97)
Common Names: Rose apple, *pomarrosa*

Family: Myrtaceae (Myrtle)

Range: Cultivated; also known from other tropical regions throughout the world.

Islands Inhabited: Floreana, Isabela (SN), San Cristóbal, Santa Cruz

Habitat: Moist uplands

Description: Tree to 15 m tall. *Leaves* opposite, simple, lanceolate-elliptic, 12–20 cm long, leathery, margins entire. *Flowers* in terminal racemes; corolla white or greenish white, petals 4, ca. 1–2 cm long; stamens numerous, white, much longer than the petals. *Fruit* a berry, whitish yellow to pinkish yellow, roundish, 3–4 cm long and 5–6 cm across; seed 1.

Comments: These trees are found near most of the towns on the four inhabited islands. Their fragrant-smelling fruits can be eaten raw, but they are best when prepared as preserves. This species is referred to as *Eugenia jambos* L. in Wiggins and Porter (1971).

97. *Syzygium jambos* (rose apple), flowers

Scientific Name: Rhizophora mangle L. (Photos 98–100)
Common Names: Red mangrove, *mangle rojo*

Family: Rhizophoraceae (Mangrove)

Range: Native; also known from other tropical regions throughout the world.

Islands Inhabited: Española, Fernandina, Floreana, Genovesa, Isabela (A,CA,D,SN), Pinzón, Rábida, San Cristóbal, Santa Cruz, Santiago, Islet(s)

Habitat: Coastal zone

Description: Tree or shrub 3–7 m tall, young branches reddish, conspicuous prop or stilt roots produced by the trunk and lower branches. *Leaves* opposite, simple, obovate to elliptic, 6.5–12.5 cm long, leathery and somewhat fleshy, margins entire and slightly rolled under. *Flowers* in axillary pairs; calyx greenish yellow, campanulate with 4 lobes, lobes 8–9 mm long, fleshy; corolla white to pale yellow, petals 4, 7–8 mm long, hairy; stamens 8. *Fruit* a berry, greenish brown, cone-shaped, 1.5–2.5 cm long; seed 1, germinating while still attached to the plant so that the brownish embryonic root and young stem extend 15–25 cm.

Comments: This plant often forms impenetrable thickets along shorelines. Species such as *Rhizophora mangle* that produce seeds with the ability to germinate while still attached to the parent are said to be "viviparous." Red mangroves often "plant" themselves when their germinated seeds fall to the ground and embed themselves upright in the moist, sandy soil.

98. *Rhizophora mangle* (red mangrove)

99. (left) *Rhizophora mangle* (red mangrove), flowers and fruit

100. (below) *Rhizophora mangle* (red mangrove), germinated seed

Scientific Name: Tectona grandis L. f. (Photo 101)
Common Names: Teak, *teca*

Family: Verbenaceae (Vervain)

Range: Introduced; also known from other tropical regions throughout the world, originally from southeast Asia.

Islands Inhabited: Santa Cruz

Habitat: Moist uplands

Description: Tree to 50 m tall, branches densely hairy. *Leaves* opposite, simple; blade broadly elliptic, 11–85 cm long, somewhat leathery, drooping, lower surface densely hairy, hairs orange-brown, margins entire or wavy and minutely toothed. *Flowers* in terminal and axillary cymes, arranged in panicles to 80 cm long, densely hairy; corolla white with purplish pink lobes, salverform with 5–7 lobes, tube 1.5–3 mm long, lobes 2.5–3 mm long; stamens 5–6. *Fruit* drupaceous, roundish, to 1.5 cm in diameter, showing 4 sections when mature, enclosed in a greatly inflated calyx, ca. 2.5 cm in diameter; seed 1 per section.

Comments: Teak is cultivated for its valuable wood, which due to its water-resistance and durability is popular in shipbuilding.

101. *Tectona grandis* (teak), flowers and fruits

See also

Chiococca alba (p. 152)
Coffea arabica (p. 153)
Scalesia helleri (p. 150)

Trees with Opposite Leaves and Yellow or Orange Flowers

Scientific Name: Spathodea campanulata P. Beauv. (Photo 102)
Common Names: African tulip tree, fountain tree, *tulipán africano*

Family: Bignoniaceae (Bignonia)

Range: Cultivated; also known from other tropical regions throughout the world, originally from tropical Africa.

Islands Inhabited: Santa Cruz

Habitat: Moist uplands

Description: Tree to 25 m tall. *Leaves* opposite or in whorls of 3, odd-pinnately compound, to 45 cm long; leaflets 3–19, elliptic, to 12 cm long. *Flowers* in terminal racemes; calyx to ca. 6 cm long, somewhat enclosing the base of the corolla, covered with reddish brown hairs; corolla reddish orange, campanulate with 5 lobes, to 12 cm long, somewhat inflated and slightly 2-lipped; stamens 4. *Fruit* a capsule, oblong-lanceolate, ca. 20 cm long; seeds numerous, winged.

Comments: Spathodea campanulata is often called the "fountain tree" because the buds, when pinched, will shoot out a stream of water. Several representatives are found on farms in the highlands.

102. *Spathodea campanulata* (African tulip tree), flowers

See also

Avicennia germinans (p. 80)
Chiococca alba (p. 152)
Rhizophora mangle (p. 86)

Trees with Opposite Leaves and Pink, Red, or Purple Flowers

Scientific Name: Syzygium malaccense (L.) Merr. & L. M. Perry
(Photos 103–4)
Common Names: Malay apple, *pera noruega*

Family: Myrtaceae (Myrtle)

Range: Cultivated; also known from other tropical regions throughout the world.

Islands Inhabited: Floreana, Santa Cruz

Habitat: Moist uplands

Description: Tree to ca. 15 m tall. *Leaves* opposite, simple, elliptic to oblong-obovate, to 35 cm long, margins entire. *Flowers* in axillary racemes, also forming on older branches and sometimes the trunk; corolla reddish purple, petals 4, 6–10 mm long; stamens numerous, reddish purple, much longer than the petals. *Fruit* a berry, reddish to maroon, obovoid to oblong, 4–7.5 cm long; seed 1.

Comments: As the flowers of this tree fall to the ground, they form a beautiful reddish carpet. The fruits can be eaten raw or cooked for preserves. This species is referred to as *Eugenia malaccencis* L. in Wiggins and Porter (1971).

103. (above) *Syzygium malaccense* (Malay apple), flowers

104. (left) *Syzygium malaccense* (Malay apple), flowers

Scientific Name: Cinchona succirubra Pav. ex Klotzsch

(Photos 105–6; see also Photo 32)

Common Names: Quinine tree, *cascarilla*

Family: Rubiaceae (Madder)

Range: Cultivated escape; also known from western South America.

Islands Inhabited: Santa Cruz

Habitat: Moist uplands

Description: Tree to ca. 12 m tall. *Leaves* opposite, simple; blade broadly ovate, ca. 10–22 cm long, upper surface shiny green at first, turning red with age, margins entire. *Flowers* in terminal panicles; corolla pinkish, salverform with 5 lobes, ca. 1.5 cm long, throat covered with whitish pink hairs; stamens 5. *Fruit* a capsule, oblong, 1–2 cm long; seeds numerous, winged.

Comments: According to Hamann (1974b), this plant was introduced into the Galápagos in 1946. No doubt, the intention was to produce quinine, a well-known compound that is obtained from the tree's bark and used to treat malaria. Ironically, malaria does not occur in the archipelago. By 1972, the tree had firmly established itself on the island. Since that time, it has become a nuisance in the highlands due to its ability to outcompete the native vegetation. It spreads at an alarming rate, due to its wind-borne seeds, and is difficult to kill once established. Currently it is advancing on the Miconia and Fern-Sedge Zones, and it may eventually dominate these areas if not successfully controlled.

105. *Cinchona succirubra* (quinine tree)

106. *Cinchona succirubra* (quinine tree), flowers

Trees with Opposite Leaves and Green Flowers

Scientific Name: Pisonia floribunda Hook. f. (Photos 107–8)
Common Names: Galápagos pisonia, *pega pega*

Family: Nyctaginaceae (Four-o'clock)

Range: Endemic

Islands Inhabited: Fernandina, Floreana, Isabela (A,CA,SN), Pinta, Pinzón, Santa Cruz, Santiago

Habitat: Arid lowlands and moist uplands

Description: Tree to 15 m tall. *Leaves* usually opposite, simple, elliptic to roundish, 3.5–8 cm long, margins entire. *Flowers* unisexual (plants dioecious), in terminal or axillary cymes. Staminate calyx yellowish green, campanulate, 5–6 mm long; corolla absent; stamens usually 6. Pistillate calyx yellowish green, somewhat tubular, 3–4 mm long; corolla absent. *Fruit* an achene, club shaped, ca. 1 cm long, 10-ribbed with sticky glands on the ribs; seed 1.

Comments: The fruits of pisonia are extremely sticky and are known to be dispersed by becoming attached to the feathers of birds. The genus name honors the seventeenth-century Dutch naturalist Wilhelm Piso.

107. (above) *Pisonia floribunda* (Galápagos pisonia), flowers

108. (left) *Pisonia floribunda* (Galápagos pisonia), fruits

Trees with Whorled Leaves and Yellow or Orange Flowers

See

Spathodea campanulata (p. 89)

Trees with Whorled Leaves and Green Flowers

Scientific Name: Casuarina equisetifolia L. (Photo 109)
Common Names: Australian pine, horsetail tree, ironwood, *arbol de hierro*

Family: Casuarinaceae (Beefwood)

Range: Introduced; also known from other tropical regions throughout the world, originally from Australia.

Islands Inhabited: Santa Cruz

Habitat: Arid lowlands

Description: Tree to ca. 20 m tall, young branchlets green, slender, and drooping. *Leaves* in whorls of ca. 7, regularly spaced along the young branchlets, simple, toothlike or scalelike, very small. *Flowers* unisexual (plants monoecious or dioecious), greenish. Staminate flowers in terminal spikes to 6 cm long, each flower with 1 bract and 2 bracteoles beneath; calyx absent; corolla absent; stamen 1. Pistillate flowers in axillary headlike clusters, each flower with 1 bract and 2 bracteoles beneath; calyx absent; corolla absent. *Fruit* a nut, flattened and winged, enclosed by the 2 bracteoles, which become woody. All fruits of each cluster fuse together into a single roundish conelike structure, ca. 1–2 cm long; seed 1.

Comments: The genus name comes from the resemblance of this tree's drooping branchlets to the feathers of the flightless cassowary bird, which belongs in the genus *Casuarinus.* These branchlets also resemble certain members of a plant group known as horsetails (genus *Equisetum*), thus the specific epithet *equisetifolia* as well as the common name "horsetail tree." This plant's resemblance to some species of the genus *Pinus* accounts for the common name "Australian pine." Although not abundant in the Galápagos Islands, this species can become weedy if not monitored. In certain parts of the world it is common along seashores, and it serves as a good windbreak.

109. *Casuarina equisetifolia* (Australian pine)

Trees with Clustered Leaves and White Flowers

See

Bursera graveolens (p. 48)
B. malacophylla (p. 50)
Carica papaya (p. 51)
Terminalia catappa (p. 52)

Trees with Clustered Leaves and Yellow or Orange Flowers

Scientific Name: Cocos nucifera L. (Photos 110–11)
Common Names: Coconut palm, *palma de coco, cocotero*

Family: Arecaceae (Palm)

Range: Cultivated escape; also known from other tropical regions throughout the world, possibly Melanesian in origin.

Islands Inhabited: Floreana, Isabela (SN), San Cristóbal, Santa Cruz

Habitat: Coastal zone

Description: Tree to 25 m tall, trunk often somewhat curved. *Leaves* arranged in a terminal cluster, pinnately compound, to 6 m long; leaflets to 1 m long. *Flowers* unisexual (plants monoecious), in axillary clusters to 1.5 m long, staminate and pistillate flowers found on the same inflorescence, with 2 large bracts beneath. Corolla of staminate flower yellow, 1 cm long, petals 3, stamens 6. Corolla of pistillate flower yellow, 2.5 cm across, petals 3. *Fruit* a large drupe composed of 3 distinct layers, the outer yellowish orange when mature, the middle thick and fibrous, the inner hard and bony, to 30 cm long; seed 1, large and hollow, containing a nutritive tissue known as "endosperm."

Comments: These trees are common in and around the coastal towns of the archipelago. Because their fruits are able to withstand long periods of emersion in seawater, coconut palms are found on tropical beaches thoughout the world. Although this mode of transportation could explain how they came to the Galápagos, it is more likely that they arrived with the early colonists. In some parts of the world the coconut fruit is extremely important to the economy. The fibrous middle layer is used to make cords and brushes, while the hard inner layer is used for making uten sils and novelty items. The fleshy endosperm is a major source of vegetable oil, and the liquid endosperm or "milk" is used as a drink and for cooking. In addition, sap from the flower clusters is fermented to produce a drink known as "toddy." The name of the genus is apparently derived from the Portuguese *coco*, meaning "monkey." This is because the inner layer of the fruit possesses 3 round marks that resemble the eyes and mouth of a monkey's face.

110. *Cocos nucifera* (coconut palm), flowers and fruits

111. *Cocos nucifera* (coconut palm), flowers and fruits

Shrubs with Alternate Leaves and White Flowers

Scientific Name: Vallesia glabra (Cav.) Link (Photos 112–13)
Common Names: Pearl berry, *peralillo, sauquillo*

Family: Apocynaceae (Dogbane)

Range: One variety is endemic, the other native.

Islands Inhabited: Española, Floreana, Isabela (A,CA,D,E,SN,W), San Cristóbal, Santa Cruz, Santiago

Habitat: Arid lowlands and moist uplands

Description: Shrub or small tree to ca. 6 m tall. *Leaves* alternate, simple; blade lanceolate to oblong-lanceolate, 2.5–9 cm long, margins entire. *Flowers* in cymes, often opposite the leaves; corolla white and greenish, salverform with 5 lobes, 5–7 mm long; stamens 5. *Fruit* a drupe, white, oblong, 1 cm long; 1–2 seeds.

Comments: This species includes two varieties. *Vallesia glabra* var. *pubescens* (Andersson) Wiggins (Photos 112–13) has stems, leaves, and inflorescences that are covered with fine hairs. This endemic is found on Española, Floreana, Isabela (A), San Cristóbal, Santa Cruz, and Santiago. *Vallesia glabra* var. *glabra,* a native, differs in that it lacks the above-mentioned hairs. With the exception of Santiago, it occurs on the same islands as var. *pubescens.* On Isabela it is known to inhabit all six volcanoes. It occurs in other parts of tropical America as well, including Mexico and the northern half of South America.

112. (above) *Vallesia glabra* var. *pubescens* (pearl berry), flowers and fruits

113. (left) *Vallesia glabra* var. *pubescens* (pearl berry), flowers and fruits

Scientific Name: Baccharis gnidiifolia Kunth (Photo 114)
Common Name: Baccharis

Family: Asteraceae (Sunflower)

Range: Native; also known from mainland Ecuador.

Islands Inhabited: Fernandina, Isabela (A,D,E,N,W), Santiago

Habitat: Moist uplands

Description: Shrub to 2 m tall, much-branched. *Leaves* alternate, simple; blade linear to linear-elliptic, 5 or more times longer (1.5–4 cm) than wide (0.5–7 mm), margins entire. *Flowers* unisexual (plants dioecious), in discoid heads ca. 2.5–3 mm across; heads in terminal or axillary clusters; flowers white. *Fruit* an achene, ca. 1–1.5 mm long, with numerous hairlike bristles on top, 2–3 mm long; seed 1.

Comments: Baccharis steetzii Andersson (Steetz's baccharis), an endemic, is found on Floreana, Isabela, and San Cristóbal. It differs from *B. gnidiifolia* in having elliptic or elliptic-ovate leaves that are only two or three times longer (2–4 cm) than wide (9–17 mm). Both species are considered rare.

114. *Baccharis gnidiifolia* (baccharis), flowers

Scientific Name: Darwiniothamnus lancifolius (Hook. f.) Harling (Photo 115)
Common Name: Lance-leafed Darwin's shrub

Family: Asteraceae (Sunflower)

Range: Endemic

Islands Inhabited: Fernandina, Isabela (A,CA,D,E,SN,W)

Habitat: Arid lowlands and moist uplands

Description: Shrub to ca. 3 m tall, much-branched. *Leaves* alternate, usually clustered near the branch tips, simple, lanceolate to broadly lanceolate, to 10 cm long, margins usually entire, occasionally minutely dentate. *Flowers* in radiate heads to ca. 1 cm across; heads solitary or in clusters, borne near the branch tips; disc flowers yellow, perfect; ray flowers white, pistillate. *Fruit* an achene, ca. 1 mm long, with numerous hairlike bristles on top, 2 mm long; seed 1.

Comments: Darwiniothamnus, which literally means "Darwin's shrub," is one of seven angiosperm genera found only in the Galápagos. According to Lawesson and Adsersen (1987), *D. lancifolius* includes three subspecies. *Darwiniothamnus lancifolius* subsp. *glabriusculus* (A. Stewart) Lawesson & Adsersen (Photo 115), referred to as *D. tenuifolius* var. *glabriusculus* (Stewart) Cronq. in Wiggins and Porter (1971), is considered rare and occurs only on the upper slopes of Volcán Sierra Negra (Isabela). *Darwiniothamnus lancifolius* subsp. *lancifolius,* also listed as *D. tenuifolius* var. *glabriusculus* in Wiggins and Porter (1971), occurs on Fernandina and Isabela (A,CA,D,SN). On Volcán Sierra Negra it typically inhabits the lower elevations. *Darwiniothamnus lancifolius* subsp. *glandulosus* Harling, referred to as *D. tenuifolius* var. *glandulosus* (Harling) Cronq. in Wiggins and Porter (1971), also is found on Fernandina and Isabela (A,CA,D,E,W). It has leaves that are somewhat glandular, while the other two subspecies generally have leaves without glands.

115. *Darwiniothamnus lancifolius* subsp. *glabriusculus* (lance-leafed Darwin's shrub), flowers

Scientific Name: Darwiniothamnus tenuifolius (Hook. f.) Harling
(Photos 116–17)
Common Name: Thin-leafed Darwin's shrub

Family: Asteraceae (Sunflower)

Range: Endemic

Islands Inhabited: Fernandina, Floreana, Isabela (A,CA,D,SN,W), Pinta, Pinzón, Santa Cruz, Santiago

Habitat: Arid lowlands and moist uplands

Description: Shrub to ca. 3 m tall, much-branched. *Leaves* alternate, usually clustered near the branch tips, simple, linear to narrowly lanceolate, to 10 cm long, margins entire. *Flowers* in radiate heads to ca. 1 cm across; heads solitary or in clusters, borne near the branch tips; disc flowers yellow, perfect; ray flowers white, pistillate. *Fruit* an achene, ca. 1 mm long, with numerous hairlike bristles on top, 2 mm long; seed 1.

Comments: Darwiniothamnus tenuifolius thrives in a variety of habitats, from lava fields to moist highlands. Its leaves have a pleasant odor when crushed, and its flowers have a mild, sweet odor. However, prolonged contact with the dried flowers and fruits has been known to cause sneezing and watery eyes in some individuals. Darwin's finches have been observed eating the flowers and fruits of this plant on Pinta.

Darwiniothamnus lancifolius (lance-leafed Darwin's shrub) is differentiated from this species primarily by its leaves, which are described as lanceolate to broadly lanceolate. *Darwiniothamnus alternifolius* Lawesson & Adsersen (alternate-leafed Darwin's shrub) differs from both in having its leaves obviously alternate and not clustered near the branch tips. In addition, its inflorescences generally are positioned well above the terminal leaves, while those of the other two species are normally located at approximately the same level as the terminal leaves (Lawesson and Adserson 1987). *Darwiniothamnus alternifolius* is found only on Volcán Cerro Azul and Volcán Sierra Negra (Isabela). It is considered a vulnerable species.

116. *Darwiniothamnus tenuifolius* (thin-leafed Darwin's shrub), flowers

117. *Darwiniothamnus tenuifolius* (thin-leafed Darwin's shrub), flowers and fruits

Scientific Name: Scalesia affinis Hook. f. (Photos 118–19)
Common Names: Radiate-headed scalesia, *lechoso, tabaquillo*

Family: Asteraceae (Sunflower)

Range: Endemic

Islands Inhabited: Fernandina, Floreana, Isabela (A,CA,D,E,SN,W), Santa Cruz

Habitat: Arid lowlands

Description. Shrub to ca. 3 m tall, rarely taller and treelike. *Leaves* alternate or almost opposite, simple, typically 6–20 cm long (including petiole); petiole sometimes winged; blade ovate to almost rhombic, both surfaces hairy and glandular, margins serrate or dentate. *Flowers* in radiate heads, disc 1 2 cm across; disc flowers white, ca. 50–100 per head, perfect; ray flowers white, 4–13 per head, usually pistillate but occasionally with a few stamens, sterile. *Fruit* an achene, flattened, to 4 mm long; seed 1.

Comments: Scalesia affinis is the only species of the genus with truly radiate heads. Specimens from Floreana and Santa Cruz normally have leaves with winged petioles, while those from Fernandina and Isabela have leaves that are unwinged or possess relatively narrow wings. This species is considered rare.

Scalesia aspera Andersson (rough-leafed scalesia) also inhabits Santa Cruz (the northwestern part). In addition, it has been recorded on the islet of Edén, but nowhere else in the archipelago. This rare plant may take the form of a shrub or small tree but typically reaches a height of ca. 1 m. It has alternate or almost opposite leaves, and its leaf blades are typically elliptic in shape and 5–10 cm long. The leaf margins are either serrate, dentate, or entire. Each discoid head is up to 2 cm across and has ca. 50–100 or more white flowers.

118. *Scalesia affinis* (radiate-headed scalesia)

119. *Scalesia affinis* (radiate-headed scalesia), flowers

Scientific Name: Scalesia baurii B. L. Rob. & Greenm. (Photos 120–21)
Common Names: Baur's scalesia, *lechoso*

Family: Asteraceae (Sunflower)

Range: Endemic

Islands Inhabited: Pinta, Pinzón, Wolf

Habitat: Arid lowlands

Description: Shrub or small tree to 3.5 m tall. *Leaves* alternate or almost opposite, simple; blade triangular to ovate, ca. 3.5–10 cm long, lobed, sinuses extending ca. one-third to three-fifths the distance to the midrib, each lobe divided again or toothed, both surfaces hairy. *Flowers* generally in discoid heads, disc to 1.5 cm across; disc flowers white, ca. 30–100 or more per head, perfect; some irregular ray flowers may also be present, white, sterile. *Fruit* an achene, flattened, to 4 mm long; seed 1.

Comments: Scalesia baurii, listed as *S. incisa* Hook. f. in Wiggins and Porter (1971), includes two subspecies. *Scalesia baurii* subsp. *baurii* occurs on Pinta, Pinzón, and Wolf. It does not possess the irregular ray flowers mentioned above, and its dried leaves appear blackish. *Scalesia baurii* subsp. *hopkinsii* (B. L. Rob.) Eliasson (Photos

120–21) does include individuals with irregular ray flowers. These unusual flowers appear to be intermediate between disc flowers and true ray flowers. The leaves of this subspecies, which is restricted to Pinta and Wolf, do not turn blackish as they dry. Both of these subspecies are considered rare. In fact, Adsersen (1989) reports that in 1974, only 10 mature specimens of subsp. *hopkinsii* were found on Pinta. However, after introduced goats were eradicated from this island, the plants began to increase in number. Observations by the author during the summers of 1990 and 1993 indicate that this subspecies appears to be making a comeback.

Scalesia retroflexa Hemsl. and *S. divisa* Andersson are two other currently accepted species whose members were originally lumped into *S. incisa* by Wiggins and Porter (1971). *Scalesia retroflexa* is known only from southeastern Santa Cruz and is considered endangered. This shrub (to ca. 2 m tall) has alternate leaves, with triangular to almost ovate blades that are usually 6–12 cm long. Additionally, the leaves are lobed, with sinuses extending only ca. one-half the distance to the midrib or less. Each lobe is divided again. The heads, generally discoid, are up to 1.5 cm across and possess ca. 30–70 disc flowers. Occasionally a few ray flowers are present. All of the flowers are white.

Scalesia divisa is restricted to San Cristóbal. This shrub or small tree, which reaches a maximum height of 4 m, has opposite or almost opposite leaves. Its leaf blades are ovate in shape, typically 3–7 cm long, and usually have dentate or entire margins. Occasionally the leaves possess toothed lobes, with sinuses extending ca. one-third the distance to the midrib. Each discoid head has ca. 30–100 or more white flowers. This rare member of the Galápagos flora may be seen at Sappho Cove.

Scalesia incisa Hook. f. (cut-leafed scalesia), as defined here, is a small shrub (to 1 m tall) with alternate or almost opposite leaves. The lobed blades are typically 4–6 cm long, and the sinuses extend from one-half to three-fourths the distance to the midrib. Each lobe is divided again or toothed. The heads of this plant are generally discoid and may be up to 1.5 cm across. Each head has ca. 30–50 disc flowers, and occasionally a few ray flowers. All of the flowers are white. This plant is known only from northern San Cristóbal and is considered rare.

120. *Scalesia baurii* subsp. *hopkinsii* (Baur's scalesia)

121. *Scalesia baurii* subsp. *hopkinsii* (Baur's scalesia), flowers

Scientific Name: Scalesia microcephala B. L. Rob. (Photos 122–23)
Common Names: Small-headed scalesia, *lechoso*

Family: Asteraceae (Sunflower)

Range: Endemic

Islands Inhabited: Fernandina, Isabela (A,D,E,W), Wolf

Habitat: Moist uplands

Description: Shrub or small tree, typically 1.5–4 m tall. *Leaves* alternate, simple; blade narrowly triangular, ovate-lanceolate, ovate, or cordate, ca. 3.5–10 cm long, both surfaces hairy, margins entire. *Flowers* in discoid heads, disc 5–7 mm across, flowers white, ca. 15–25 per head, perfect. *Fruit* an achene, flattened, ca. 3.5 mm long, occasionally with a few small bumps on top or 1–2 awns up to 1.5 mm long; seed 1.

Comments: This species includes two varieties. *Scalesia microcephala* var. *cordifolia* Eliasson has cordate leaf bases and occurs on Isabela (E,W) and Wolf. *Scalesia microcephala* var. *microcephala* (Photos 122–23) has obtuse, rounded, or squared leaf bases and is an important component of the highland forests on Fernandina and Isabela (A,D).

Scalesia cordata A. Stewart (heart-leafed scalesia) (Photo 124) is a tree that can grow to over 10 m tall. It possesses alternate leaves, with cordate-shaped blades that are typically 4–9 cm long. Each discoid head is 5–7 mm across and has ca. 15–30 or more white flowers. This species has inflorescences that resemble those of *S. microcephala* in general, and leaves that are similar to those of *S. microcephala* var. *cordifolia* in particular. However, the leaves of *S. cordata* are more strongly heart-shaped in outline and often have toothed margins. The achenes also set these two species apart. Those of *S. cordata* almost always have two awns on top that may reach 2 mm in length. There may be several smaller scales between these awns as well. This rare species is known only from Isabela (CA,SN).

122. *Scalesia microcephala* var. *microcephala* (small-headed scalesia)

123. *Scalesia microcephala* var. *microcephala* (small-headed scalesia), flowers

124. *Scalesia cordata* (heart-leafed scalesia)

Scientific Name: Scalesia stewartii Riley (Photos 125–26)
Common Names: Stewart's scalesia, *lechoso*

Family: Asteraceae (Sunflower)

Range: Endemic

Islands Inhabited: Santiago, Islet(s)

Habitat: Arid lowlands

Description: Shrub or small tree to 3 m tall. *Leaves* alternate, simple; blade ovate to ovate-lanceolate, typically 6–10 cm long, lower surface with long, white hairs, margins entire. *Flowers* in discoid heads, disc to 1.5 cm across, flowers white, ca. 35–90 per head, perfect. *Fruit* an achene, flattened, to 4 mm long; seed 1.

Comments: This species is named for Alban Stewart, who spent almost a year collecting Galápagos plants in 1905–6. It is easily observed near the beach on Bartolomé and may also be found at nearby Sullivan Bay. Its status is considered vulnerable.

125. *Scalesia stewartii* (Stewart's scalesia)

126. *Scalesia stewartii* (Stewart's scalesia), flowers

Scientific Name: Scalesia villosa A. Stewart (Photo 127)
Common Names: Longhaired scalesia, *lechoso*

Family: Asteraceae (Sunflower)

Range: Endemic

Islands Inhabited: Floreana, Islet(s)

Habitat: Arid lowlands

Description: Shrub or small tree to 3 m tall. *Leaves* alternate, simple; blade lanceolate, ca. 6–11 cm long, both surfaces covered with long, white nonglandular hairs and short glandular hairs, margins entire and often rolled under. *Flowers* in discoid heads, disc ca. 2 cm across, flowers white, ca. 200–300 per head, perfect. *Fruit* an achene, flattened, 3–4 mm long; seed 1.

Comments: This plant, considered rare, is easily identified by its extreme hairiness. It is one of the more obvious components of the vegetation at Punta Cormorán on Floreana, where it can be seen with *Lecocarpus pinnatifidus* (wing-fruited lecocarpus).

127. *Scalesia villosa* (longhaired scalesia), fruits

Scientific Name: Cordia leucophlyctis Hook. f. (Photo 128; see also Photo 16)
Common Name: Cordia

Family: Boraginaceae (Borage)

Range: Endemic

Islands Inhabited: Española, Fernandina, Isabela (A,CA,D,SN,W), Pinzón, Santa Cruz, Santa Fe, Santiago

Habitat: Arid lowlands and moist uplands

Description: Shrub to 2.5 m tall, young branches hairy. *Leaves* alternate, simple; blade lanceolate, to ca. 4 cm long, upper surface rough, covered with upright hairs, margins serrate to dentate. *Flowers* in compact clusters or short spikes; corolla white, funnelform with 5 lobes, 5–10 mm long; stamens 5. *Fruit* a drupe, bright red at maturity, ovoid, 4–6 mm long; seeds 1–4.

Comments: Cordia anderssonii (Kuntze) Gürke (Andersson's cordia) and *C. scouleri* Hook. f. (Scouler's cordia), both endemics, closely resemble *C. leucophlyctis* but may be distinguished as follows: The flowers of *C. anderssonii* have calyx lobes that are covered with longer hairs than those found at the base of the calyx, while *C. leucophlyctis* and *C. scouleri* produce hairs of equal length. The latter two species differ in that *C. leucophlyctis* possesses simple, upright hairs on its stems and the lower surface of its leaves, whereas *C. scouleri* has both simple, upright hairs and tiny stellate hairs on these parts. *Cordia anderssonii* is found on Española, Isabela (A,D, SN), Pinta, Pinzón, San Cristóbal, Santa Cruz, and Santiago, while *C. scouleri* inhabits Floreana, Pinta, Santa Cruz, and Santiago.

128. *Cordia leucophlyctis* (cordia), flowers

Scientific Name: Cordia revoluta Hook. f. (Photos 129–30)
Common Names: Revolute-leafed cordia, *laurelillo*

Family: Boraginaceae (Borage)

Range: Endemic

Island's Inhabited: Fernandina, Floreana, Isabela (A,CA,D,SN,W), Santiago

Habitat: Arid lowlands and moist uplands

Description: Shrub 2–4 m tall, possessing many slender, upright branches. *Leaves* alternate, simple; blade linear, 4–7 cm long, upper surface rough, covered with flattened hairs, margins entire, rolled under. *Flowers* in compact clusters; corolla white, tubular with 5 lobes, 7–12 mm long; stamens 4–5. *Fruit* a drupe, bright red at maturity, ovoid, 5–6 mm long; seeds 1–4.

Comments: With its combination of linear leaves and tubular flowers, this beautiful shrub should not be confused with any other species in the genus. Inhabiting a wide variety of localities, from barren lava to forests, it is perhaps best seen on the lava fields near the town of Puerto Villamil (Isabela).

Cordia polycephala (Lam.) I. M. Johnst., a native, is considered rare. However, it has been reported from the south slope of Sierra Negra (Isabela) between 200–600 m altitude.

129. (below) *Cordia revoluta* (revolute-leafed cordia)

130. (right) *Cordia revoluta* (revolute-leafed cordia), flowers and fruit

Scientific Name: Tournefortia pubescens Hook. f. (Photo 131)
Common Name: White-haired tournefortia

Family: Boraginaceae (Borage)

Range: Endemic

Islands Inhabited: Fernandina, Floreana, Isabela (A,CA,D,SN,W), Pinzón, San Cristóbal, Santa Cruz, Santiago, Wolf

Habitat: Arid lowlands and moist uplands

Description: Shrub to 4 m tall, younger branches with mostly whitish hairs. *Leaves* alternate, simple; blade usually elliptic or ovate, to 15 cm long, lower surface with whitish hairs, margins entire. *Flowers* in axillary and terminal, branching, scorpioid cymes; corolla white, often with a greenish yellow throat, tubular with 5 lobes, ca. 4–5 mm long, 4 mm across; stamens 5. *Fruit* a drupe consisting of 2–4 nutlets, white, roundish, 3–6 mm in diameter; seed 1 per nutlet.

Comments: Tournefortia pubescens is often misidentified as *T. rufo-sericea* (rufous-haired tournefortia). However, *T. rufo-sericea* has distinct reddish brown hairs on its young branches rather than whitish hairs. Another difference involves the calyx lobes. Those of *T. pubescens* extend only halfway to the base of the calyx, while those of *T. rufo-sericea* may extend all the way to the base of the calyx.

131. *Tournefortia pubescens* (white-haired tournefortia), flowers

Scientific Name: Tournefortia rufo-sericea Hook. f. (Photos 132–33)
Common Name: Rufous-haired tournefortia

Family: Boraginaceae (Borage)

Range: Endemic

Islands Inhabited: Fernandina, Floreana, Isabela (A,CA,D,SN,W), Pinta, San Cristóbal, Santa Cruz, Santiago

Habitat: Arid lowlands and moist uplands

Description: Shrub to 5 m tall, older branches reddish brown, younger branches with reddish brown hairs. *Leaves* alternate, simple; blade broadly elliptic, to 25 cm long, lower surface with reddish brown hairs, margins entire. *Flowers* in terminal, branching, scorpioid cymes with reddish brown hairs; corolla white, often with a greenish yellow throat, tubular with 5 lobes, ca. 4–6 mm long, 4 mm across; stamens 5. *Fruit* a drupe consisting of 2–4 nutlets, white, roundish, 5–6 mm in diameter; seed 1 per nutlet.

Comments: The specific epithet, *rufo-sericea,* refers to the reddish brown hairs on this plant's stems and leaves. Its fruits are a favorite food of the Galápagos mockingbird (*Nesomimus parvulus*) and certain finch species. In addition, its flowers provide nectar for a variety of insects, especially after the sun has gone down.

132. *Tournefortia rufo-sericea* (rufous-haired tournefortia), flowers

133. *Tournefortia rufo-sericea* (rufous-haired tournefortia), flowers and fruits

Scientific Name: Scaevola plumieri (L.) Vahl (Photo 134)
Common Names: Inkberry, sea grape

Family: Goodeniaceae (Goodenia)

Range: Native; also known from other tropical regions throughout the world.

Islands Inhabited: Floreana, Isabela (SN), San Cristóbal, Santa Cruz

Habitat: Coastal zone

Description: Shrub to ca. 1 m tall, much-branched. *Leaves* alternate, simple; petiole winged; blade obovate, 3–8 cm long, fleshy, margins entire. *Flowers* in cymes; corolla white, tubular with 1 distinct lip, 2–2.5 cm long, lip 5-lobed, lobes 1–1.4 cm long, throat yellowish and extremely hairy; stamens 5. *Fruit* a drupe, green at first but turning dark purple or black, roundish, ca. 1–2 cm long, containing a yellowish flesh; seed 1.

Comments: In the Galápagos, inkberry is considered rare. However, its fruits are well adapted to dispersal by ocean currents, and for this reason the species is found on beaches throughout the world. On the mainland, it is sometimes planted intentionally to reduce erosion along coastlines. The origin of the name "sea grape" is obvious, as the mature fruits do resemble grapes.

134. *Scaevola plumieri* (inkberry), flower and fruit

Scientific Name: Cryptocarpus pyriformis Kunth (Photos 135–36)
Common Names: Salt bush, *monte salado*

Family: Nyctaginaceae (Four-o'clock)

Range: Native; also known from mainland Ecuador and Peru.

Islands Inhabited: Española, Fernandina, Floreana, Genovesa, Isabela (CA,D,SN), Marchena, Pinta, Pinzón, Rábida, San Cristóbal, Santa Cruz, Santa Fe, Santiago, Seymour, Islet(s)

Habitat: Coastal zone and arid lowlands

Description: Prostrate to erect shrub, stems much-branched and several m long, covered with short sticky hairs. *Leaves* alternate, simple, ovate to somewhat cordate, 3–5 cm long, margins entire. *Flowers* in terminal and axillary panicles, tightly clustered; calyx white and greenish, campanulate with 5 lobes, ca. 2 mm long, covered with fine sticky hairs; corolla absent; stamens usually 5. *Fruit* an achene, somewhat pear-shaped with 5 lobes, ca. 1.5 mm long; seed 1.

Comments: The genus name, derived from the Greek *krypto,* "to hide," and *carpus,* "fruit," refers to the small and thus seemingly hidden fruits produced by the plant. As implied by the common names, this species is extremely salt-tolerant. Visitors will become quite familiar with salt bush, as it is frequently encountered.

135. *Cryptocarpus pyriformis* (salt bush), flowers

136. *Cryptocarpus pyriformis* (salt bush), flowers

Scientific Name: Capraria biflora L. (Photo 137)
Common Name: Hairy capraria

Family: Scrophulariaceae (Figwort)

Range: Native; also known from other tropical regions throughout the world.

Islands Inhabited: Floreana, San Cristóbal, Santa Cruz

Habitat: Arid lowlands

Description: Shrub to ca. 1 m tall, stems covered with hairs. *Leaves* alternate, simple, oblanceolate, 1.5–6 cm long, broadest above the middle, somewhat blunt at the tip, both surfaces hairy, glandular-dotted, margins dentate-serrate. *Flowers* in axillary pairs; corolla white with pale lavender lines or occasionally pinkish, campanulate, ca. 3.5–4.5 mm long, somewhat 2-lipped, upper lip 2-lobed, lower lip 3-lobed, lobes ca. 4 mm long, inside of lower lip hairy; stamens 4, anthers bluish. *Fruit* a capsule, brown, ovoid, 4–6 mm long, covered with glandular dots; seeds numerous.

Comments: Capraria biflora differs from *C. peruviana* (smooth capraria) primarily in the shape and surface covering of its leaves.

137. *Capraria biflora* (hairy capraria), flowers

Scientific Name: Capraria peruviana Benth. (Photo 138)
Common Name: Smooth capraria

Family: Scrophulariaceae (Figwort)

Range: Native; also known from Colombia, mainland Ecuador, and Peru.

Islands Inhabited: Floreana, Isabela (SN), San Cristóbal, Santa Cruz, Santiago

Habitat: Arid lowlands and moist uplands

Description: Shrub 1–2 m tall, stems smooth. *Leaves* alternate, simple, narrowly elliptic-lanceolate to linear-lanceolate, 3–11 cm long, broadest at the middle or slightly below, pointed at the tip, both surfaces smooth, glandular-dotted, margins minutely serrate. *Flowers* axillary, solitary or in clusters of 2–3; corolla greenish white to pale greenish yellow, campanulate with 5 lobes, 5–6 mm long, lobes 3–3.5 mm long; stamens 4, anthers grayish. *Fruit* a capsule, brown, ovoid, 5–7 mm long, covered with glandular dots; seeds numerous.

Comments: This plant's common name refers to the lack of hairs on its leaves and inflorescences.

138. *Capraria peruviana* (smooth capraria), flowers and fruits

Scientific Name: Acnistus ellipticus Hook. f. (Photos 139–40)
Common Names: Galápagos acnistus, *cogojo*

Family: Solanaceae (Potato)

Range: Endemic

Islands Inhabited: Fernandina, Floreana, Isabela (A,SN), Pinzón, San Cristóbal, Santa Cruz, Santiago

Habitat: Moist uplands

Description: Shrub or small tree 2–5 m tall. *Leaves* alternate, simple; petiole somewhat winged; blade elliptic to obovate, 3–20 cm long, margins entire. *Flowers* axillary, solitary or in clusters of 2–8; corolla white, often tinged with green, funnelform with 5 lobes, 2.5–3 cm long, lobes 8–10 mm long, tips often curling backward, somewhat hairy on margins and outside; stamens 5. *Fruit* a berry, black at maturity, roundish, usually ca. 1 cm in diameter; seeds numerous.

Comments: Acnistus ellipticus typically inhabits moist, forested areas. Visitors may see it while exploring the highlands of Santa Cruz.

139. *Acnistus ellipticus* (Galápagos acnistus), flowers and fruits

140. *Acnistus ellipticus* (Galápagos acnistus), flowers

Scientific Name: Brugmansia candida Pers. (Photo 141)
Common Names: Angel's trumpet, *campana, floripondio*

Family: Solanaceae (Potato)

Range: Cultivated escape; also known from other tropical regions throughout the world, originally from Peru.

Islands Inhabited: Floreana, Isabela (SN), Santa Cruz

Habitat: Moist uplands

Description: Shrub or small tree to ca. 10 m tall, stems covered with soft hairs. *Leaves* alternate, simple; blade ovate to oblong, 10–25 cm long, both surfaces hairy, margins entire. *Flowers* hanging in forks of branches, solitary; corolla white with greenish veins, funnelform with 5 lobes, to ca. 20 cm long, lobes 2.5–5 cm long; stamens 5. *Fruit* a capsule, ovoid, 4–8 cm long; seeds numerous.

Comments: This species, listed as *Datura arborea* L. in Wiggins and Porter (1971), produces flowers that are quite fragrant. The leaves, however, contain the alkaloid hyoscine and are poisonous. The genus was named in honor of the Dutch botanist S. J. Brugmans (1763–1819).

141. *Brugmansia candida* (angel's trumpet), flower

Scientific Name: Grabowskia boerhaaviaefolia (L. f.) Schltdl.
(Photos 142–43)
Common Name: Grabowskia

Family: Solanaceae (Potato)

Range: Native; also known from Peru.

Islands Inhabited: Española, Rábida, Santa Cruz, Santa Fe, Seymour, Islet(s)

Habitat: Arid lowlands

Description: Shrub to 2.5 m tall, younger branches with whitish bark and axillary spines (usually 2–3 cm long), older branches reddish brown. *Leaves* alternate or in clusters of up to 15 or more, simple, blade elliptic to ovate, 2–3 cm long, margins entire. *Flowers* axillary, solitary or in small clusters; corolla white, often tinged with purple or with purplish veins, funnelform with 5 lobes, tube 3–4 mm long, lobes 4–5 mm long; stamens 5. *Fruit* dark blue with a waxy covering, ovoid, 6–8 mm long, consisting of 4 nutlets; seeds 1–2 per nutlet.

Comments: A prime location for observing this species is Plaza Sur. One of the interesting characteristics of this plant is that its branches are frequently covered with lichens.

142. *Grabowskia boerhaaviaefolia* (grabowskia)

143. *Grabowskia boerhaaviaefolia* (grabowskia), flowers

Scientific Name: Solanum erianthum D. Don (Photo 144)
Common Name: Velvet nightshade

Family: Solanaceae (Potato)

Range: Native; also known from other tropical regions throughout the world.

Islands Inhabited: Fernandina, Floreana, Isabela (SN), Pinta, Santa Cruz, Santiago

Habitat: Arid lowlands and moist uplands

Description: Shrub or small tree 2–8 m tall, stem tips with a granular, hairy surface covering, some of the hairs stellate. *Leaves* alternate, simple, ovate or elliptic, 12–37 cm long, both surfaces somewhat velvety, margins entire or somewhat wavy. *Flowers* in lateral cymes; corolla white with green veins, rotate with 5 lobes, typically 1.1–1.8 cm across; stamens 5, anthers yellow. *Fruit* a berry, yellow, round, 1–1.2 cm in diameter, fleshy, covered with hairs; seeds numerous.

Comments: Velvet nightshade occupies both open and forested areas. Unfortunately, it does not typically grow in locations frequented by visitors.

144. *Solanum erianthum* (velvet nightshade), flowers

Scientific Name: Solanum quitoense Lam. (Photo 145)
Common Names: Purple solanum, *naranjilla*

Family: Solanaceae (Potato)

Range: Cultivated escape; also known from Venezuela, Bolivia, and mainland Ecuador.

Islands Inhabited: Isabela (SN), San Cristóbal, Santa Cruz, Santiago

Habitat: Moist uplands

Description: Shrub 1–2 m tall, stem tips appearing woolly with minute stellate hairs. *Leaves* alternate, simple, ovate, 27–34 cm long, both surfaces covered with minute stellate hairs, margins angular-wavy. *Flowers* in lateral cymes; corolla white, rotate with 5 lobes, ca. 2 cm across; stamens 5, anthers yellow. *Fruit* a berry, orange, round, ca. 5 cm in diameter, fleshy, covered with hairs at first; seeds numerous.

Comments: This species has stems, leaves, and inflorescences that are often covered with purple hairs, a trait that easily distinguishes it from other members of the genus. It is cultivated for the popular juice that is squeezed from its fruits. Visitors might happen upon this species at one of the local farms.

145. *Solanum quitoense* (purple solanum), flower and fruits

See also

Bursera graveolens (p. 48)
B. malacophylla (p. 50)
Capsicum frutescens (p. 194)
C. galapagoense (p. 194)
Carica papaya (p. 51)
Desmanthus virgatus (p. 186)
Hibiscus diversifolius (p. 136)
Leucaena leucocephala (p. 66)
Piscidia carthagenensis (p. 55)
Scalesia helleri (p. 150)

Shrubs with Alternate Leaves and Yellow or Orange Flowers

Scientific Name: Pleuropetalum darwinii Hook. f. (Photo 146)
Common Name: Galápagos pleuropetalum

Family: Amaranthaceae (Amaranth)

Range: Endemic

Islands Inhabited: Isabela (A,SN), Santa Cruz, Santiago

Habitat: Moist uplands

Description: Shrub 1–2 m tall. *Leaves* alternate, simple, elliptic to ovate, 5–10 cm long, margins entire. *Flowers* in terminal panicles, each flower with two bracteoles beneath; calyx orange, sepals 5, 4–5 mm long; corolla absent; stamens ca. 8. *Fruit* a berrylike capsule, black, roundish, ca. 6 mm in diameter, opening by means of a caplike lid; seeds numerous, black, shiny.

Comments: This shrub is considered rare. However, with determination one might find it in the highlands of one of the islands it inhabits.

146. *Pleuropetalum darwinii* (Galápagos pleuropetalum), fruits

Scientific Name: Encelia hispida Andersson (Photo 147)
Common Name: Galápagos encelia

Family: Asteraceae (Sunflower)

Range: Endemic

Islands Inhabited: Floreana, Isabela (SN), San Cristóbal, Santa Fe, Santiago

Habitat: Arid lowlands

Description: Low shrub to ca. 1 m tall, much-branched, young branches with grayish hairs. *Leaves* alternate, simple, blade lance-ovate to somewhat elliptic, 1–5 cm long, both surfaces with grayish hairs, margins typically entire. *Flowers* in radiate heads, disc to ca. 1 cm across; heads solitary or several grouped together, borne near the branch tips; disc flowers purplish, perfect; ray flowers yellow, lacking stamens and pistil. *Fruit* an achene, 4–5 mm long, covered with white hairs; seed 1.

Comments: This species is classified as rare. However, a nice population exists at Punta Pitt (San Cristóbal).

147. *Encelia hispida* (Galápagos encelia), flowers

Scientific Name: Caesalpinia bonduc (L.) Roxb. (Photos 148–49)
Common Names: Prickly caesalpinia, *mora*

Family: Caesalpiniaceae (Caesalpinia)

Range: Native; also known from other tropical regions throughout the world.

Islands Inhabited: Isabela (SN), Santa Cruz

Habitat: Coastal zone and moist uplands

Description: Low spreading shrub to 2 m tall, branches covered with prickles. *Leaves* alternate, bipinnately compound with 5–7 pairs of pinnae; rachis and pinnae with prickles; leaflets 6–8 pairs per pinna, oval to elliptic-oblong, to ca. 5 cm long. *Flowers* in axillary racemes 15–30 cm long, prickles often present; corolla yellow with one petal splotched reddish brown, 1–2 cm across, petals 5; stamens 10. *Fruit* a legume, oblong, 5–7 cm long, 3–5 cm wide, covered with prickles; seeds 1–2, gray, roundish, 1–2 cm in diameter.

Comments: This prickly shrub, with fruits adapted for dispersal by ocean currents, is often found on tropical beaches throughout the world. However, on Santa Cruz it is common in the highlands, where it appears to favor open, disturbed areas. Its occurrence here is probably a result of human introduction. One must be sure to avoid its prickles, as they can be extremely unpleasant. The genus was named for the Italian botanist Andreas Cesalpino (1519–1603).

148. (below) *Caesalpinia bonduc* (prickly caesalpinia), flowers

149. (right) *Caesalpinia bonduc* (prickly caesalpinia), flower

Scientific Name: Senna pistaciifolia (Kunth) Irwin & Barneby (Photo 150)

Common Name: Flat-fruited senna

Family: Caesalpiniaceae (Caesalpinia)

Range: Native; also known from mainland Ecuador and Peru.

Islands Inhabited: Fernandina, Floreana, Isabela (A,CA,D,SN), San Cristóbal, Santa Cruz, Santiago

Habitat: Arid lowlands

Description: Low shrub 1–2 m tall. *Leaves* alternate, even-pinnately compound; leaflets 10–14, oval to oblong, tips somewhat rounded, middle ones 5–8 cm long. *Flowers* in axillary or terminal racemes; corolla yellow, petals 5, ca. 1.5–2.0 cm long; fertile stamens 6 (2 large, 4 small). *Fruit* a legume, flattened and oblong, 7–11 cm long; seeds numerous.

Comments: All Galápagos members of this species represent var. *picta* (G. Don) Irwin & Barneby. This plant, referred to as *Cassia picta* G. Don in Wiggins and Porter (1971), typically inhabits somewhat disturbed areas such as roadsides. Its flowers are often visited by the Galápagos carpenter bee (*Xylocopa darwini*).

Senna bicapsularis (L.) Roxb. var. *bicapsularis*, listed as *Cassia bicapsularis* L. in Wiggins and Porter (1971), is an introduced weed found only on Santa Cruz. It differs from *S. pistaciifolia* var. *picta* in that it typically has 6–8 leaflets, and its fruits are cylindrical and somewhat swollen rather than flattened.

Another introduced member of this genus, also known from Santa Cruz, is *S. alata* (L.) Roxb. (candle senna) (Photo 151). It is common in and around the town of Puerto Ayora and may be identified by its large leaves and large, winged fruits.

150. *Senna pistaciifolia* var. *picta* (flat-fruited senna), flowers and fruits

151. *Senna alata* (candle senna), flowers and fruits

Scientific Name: Gossypium darwinii Watt (Photo 152)
Common Names: Darwin's cotton, *algodón de Darwin, algodoncillo*

Family: Malvaceae (Mallow)

Range: Endemic

Islands Inhabited: Española, Fernandina, Floreana, Isabela (A,CA,D,SN), Marchena, Pinta, Pinzón, Rábida, San Cristóbal, Santa Cruz, Santa Fe, Seymour, Islet(s)

Habitat: Arid lowlands and moist uplands

Description: Shrub 1–3 m tall, branches covered with black dots and various hairs. *Leaves* alternate, simple; blade usually with 3 or 5 lobes, occasionally unlobed and ovate, to 15 cm long and 20 cm wide, margins entire. *Flowers* axillary, solitary; corolla yellow with reddish purple at the base of each petal, turning pinkish orange with age, petals 5, 4–8 cm long; stamens numerous, united in a tubular column. *Fruit* a capsule, ovoid, somewhat triangular in cross section, 1.8–2.8 cm long, usually with 3 sections, covered with black pits; seeds numerous, covered with long white lint, brownish fuzz underneath.

Comments: Darwin's cotton, referred to as *Gossypium barbadense* L. var. *darwinii* (Watt.) J. B. Hutch. in Wiggins and Porter (1971), often grows in relatively open rocky areas. The other endemic member of the genus, *G. klotzschianum* Andersson (Galápagos cotton), is considered rare. It is known from Floreana, Isabela (A,D), Marchena, San Cristóbal, and Santa Cruz; it differs from Darwin's cotton in that its leaves are usually unlobed, and its seeds lack the long white lint. Neither of these species is cultivated for commercial use.

152. *Gossypium darwinii* (Darwin's cotton), flower and fruits

Scientific Name: Hibiscus tiliaceus L. (Photo 153)
Common Names: Seaside hibiscus, *mahoe, majagua*

Family: Malvaceae (Mallow)

Range: Native; also known from other tropical regions throughout the world.

Islands Inhabited: Floreana, Isabela (SN), Santa Cruz

Habitat: Coastal zone and arid lowlands

Description: Shrub or small tree typically 2–6 m tall, occasionally taller, the young branches smooth or with fine hairs. *Leaves* alternate, simple, ovate to roundish, usually 8–20 cm long, somewhat leathery, upper surface glossy, lower surface covered with white fuzzy hairs, margins entire or minutely crenate. *Flowers* axillary, solitary or in few-flowered cymes; corolla yellow, sometimes reddish at the base, turning orange and then deep red with age, petals 5, 4–8.5 cm long; stamens numerous, united in a tubular column. *Fruit* a capsule, ovoid, 1.5–3 cm long, with 5 sections (rarely 10), short-beaked; seeds numerous, reddish brown.

Comments: The level of Galápagos carpenter bee (*Xylocopa darwini*) activity at this plant's flowers is at times remarkable. While probing for the abundant nectar, a bee will work its way around the staminal column, becoming completely dusted with pollen. If the bee then moves to another plant, the chances are good that cross-pollination will occur. In this way, both the plant and the bee benefit from their relationship. Carpenter bees also make use of this species by building nests in its wood. Although of little commercial value in the Galápagos, this plant does possess fibers that are used for cordage in other parts of the world.

153. *Hibiscus tiliaceus* (seaside hibiscus), flowers and Galápagos carpenter bee (*Xylocopa darwini*)

Scientific Name: Scutia spicata (Humb. & Bonpl. ex Schult.) Weberb.

(Photos 154–55)

Common Names: Thorn shrub, *espino*

Family: Rhamnaceae (Buckthorn)

Range: Endemic

Islands Inhabited: Española, Floreana, Isabela (CA,D,SN), Rábida, San Cristóbal, Santa Cruz, Santa Fe, Santiago, Islet(s)

Habitat: Arid lowlands

Description: Shrub 0.5–2.5 m tall, much-branched, stems with brown-tipped spines 1–6 cm long. *Leaves* usually alternate, simple, mainly elliptic to oblong, 0.6–4.5 cm long, leathery, short-lived, margins entire. *Flowers* axillary, solitary or in clusters of 2–4; calyx greenish yellow, cup shaped with 5 lobes, lobes 0.8–1 mm long; corolla greenish yellow, petals 5, each somewhat hood-shaped and 2-lobed, 0.6–0.7 mm long; stamens 5. *Fruit* consisting of 2–3 nutlets, bright red to purple and fleshy at maturity, obovate, ca. 5 mm long; seeds 2–3.

Comments: This species, listed as *Scutia pauciflora* (Hook. f.) Weberb. in Wiggins and Porter (1971), is extremely common near the shore and at lower elevations. Visitors will want to take care, as there is a painful price to pay for walking too close to these spiny plants. On a positive note, the mature fruits are known to be a source of food for land iguanas on Plaza Sur and for various birds on Santa Cruz. All Galápagos members of this species represent var. *pauciflora* (Hook. f.) M. C. Johnst.

154. *Scutia spicata* var. *pauciflora* (thorn shrub), flowers

155. *Scutia spicata* var. *pauciflora* (thorn shrub), fruits

Scientific Name: Waltheria ovata Cav. (Photos 156–57)
Common Name: Waltheria

Family: Sterculiaceae (Sterculia)

Range: Native; also known from Peru.

Islands Inhabited: Española, Fernandina, Floreana, Genovesa, Isabela (A,CA,D,SN), Marchena, Pinta, Rábida, San Cristóbal, Santa Cruz, Santa Fe, Santiago, Islet(s)

Habitat: Arid lowlands

Description: Shrub 0.5–2 m tall, much-branched, younger stems covered with minute stellate hairs. *Leaves* alternate, simple; blade ovate or slightly 3-lobed, 1–6 cm long, both surfaces covered with grayish, minute stellate hairs, margins serrate. *Flowers* in terminal and axillary clusters; corolla yellow, petals 5, 6–7 mm long; stamens 5. *Fruit* a capsule, roundish, 2–3 mm long; seed 1.

Comments: This shrub is extremely common in the rocky lowlands of many of the islands. Its flowers, though small, appear to be a favorite of the Galápagos carpenter bee (*Xylocopa darwini*) and *Polistes versicolor*, a recently introduced wasp.

156. *Waltheria ovata* (waltheria), flowers

157. *Waltheria ovata* (waltheria), flowers

Scientific Name: Triumfetta semitriloba Jacq. (Photo 158)
Common Name: Triumfetta

Family: Tiliaceae (Linden)

Range: Native; also known from other regions of tropical America.

Islands Inhabited: Fernandina, Floreana, Isabela (A,CA,SN), Santa Cruz

Habitat: Arid lowlands and moist uplands

Description: Shrub 0.5–3 m tall, stems and inflorescences covered with stellate hairs. *Leaves* alternate, simple, ovate to lance-ovate, larger ones slightly 3-lobed, 3–10 cm long, margins serrate. *Flowers* in clusters opposite the leaves; corolla yellow, petals 5, ca. 5–7 mm long; stamens numerous. *Fruit* a capsule, brown, roundish, covered with short-hooked spines, 6–9 mm in diameter including spines, spines 2–3 mm long; seeds up to 10.

Comments: This shrub is relatively widespread on the islands it inhabits, and it is common in the forests, fields, and roadsides near Santo Tomás (Isabela). This, no doubt, is the result of its spiny fruits, which are ideal for sticking to any animals that pass by. Indeed, hikers often find these fruits attached to their socks and boot strings. The genus was named in honor of the Italian botanist G. B. Triumfetti (1658–1708). The specific epithet, *semitriloba*, refers to this plant's slightly 3-lobed leaves.

158. *Triumfetta semitriloba* (triumfetta), flowers and fruits

See also

Acacia rorudiana (p. 68)
Caesalpinia pulcherrima (p. 133)
Capraria peruviana (p. 117)
Castela galapageia (p. 138)
Codiaeum variegatum (p. 54)
Cordia lutea (p. 61)
Croton scouleri (p. 53)
Darwiniothamnus alternifolius (p. 102)
D. lancifolius (p. 101)
D. tenuifolius (p. 102)
Hibiscus diversifolius (p. 136)
Parkinsonia aculeata (p. 62)
Prosopis juliflora (p. 69)
Trema micrantha (p. 70)
Zanthoxylum fagara (p. 70)

Shrubs with Alternate Leaves and Pink, Red, or Purple Flowers

Scientific Name: Caesalpinia pulcherrima (L.) Sw. (Photo 159)
Common Name: Dwarf poinciana

Family: Caesalpiniaceae (Caesalpinia)

Range: Introduced; also known from other tropical regions throughout the world.

Islands Inhabited: Santa Cruz

Habitat: Arid lowlands

Description: Shrub to ca. 4 m tall, branches with prickles. *Leaves* alternate, bipinnately compound with 6–9 pairs of pinnae, leaflets 10–12 pairs per pinna, oblong, to 2.5 cm long. *Flowers* in terminal or axillary racemes; corolla orange-red with yellow on the margins of the petals, to ca. 5 cm across, petals 5, stamens 10. *Fruit* a legume, broadly linear, to 10 cm long; seeds numerous.

Comments: Dwarf poinciana is one of the showiest members of the Galápagos flora. It is commonly seen around dwellings, and the flowers are often used as decorations in restaurants. The beautiful red and yellow petals are accented by stamens consisting of long red filaments with yellow anthers. The petals will turn completely red as they begin to age.

159. *Caesalpinia pulcherrima* (dwarf poinciana), flowers and fruits

Scientific Name: Euphorbia milii Des Moul. (Photo 160)
Common Name: Crown-of-thorns

Family: Euphorbiaceae (Spurge)

Range: Cultivated; also known from other tropical regions throughout the world, originally from Madagascar.

Islands Inhabited: Santa Cruz

Habitat: Arid lowlands

Description: Shrub to ca. 1.5 m tall, somewhat climbing, containing a milky sap, branches covered with spines. *Leaves* alternate, crowded near the branch tips, simple, obovate to spatulate, to ca. 6 cm long, margins entire. *Flowers* unisexual (plants monoecious), in clusters of cup-shaped structures known as "cyathia." Each cluster has 2 bright red, broadly ovate bracts beneath, ca. 1.5 cm across. Staminate flowers numerous in each cyathium; calyx absent; corolla absent; stamen 1. Pistillate flowers 1 per cyathium; calyx absent; corolla absent. *Fruit* a capsule; seed 1.

Comments: This plant's specific epithet honors Baron Milius, who introduced it to France in 1821. The common name, of course, refers to the thorny crown worn by Christ during his crucifixion. The milky sap produced by this shrub has been known to cause burning and blistering of the skin, and even temporary blindness, and death may result if any of its parts are eaten. All of these problems can be avoided if care is taken when handling the plant—a small inconvenience, considering the beauty of its "flowers."

Euphorbia tirucalli L. (pencil tree) (Photo 161) is another cultivated member of this genus that visitors may encounter on Santa Cruz. Its common name refers to the fleshy, green stems, which are somewhat pencil-shaped.

Only one member of the genus is endemic to the archipelago. This is *E. equisetiformis* A. Stewart (Galápagos spurge). It is also the only one that grows in the wild, occurring on Volcán Cerro Azul and Volcán Sierra Negra (Isabela). Visitors rarely see this much-branched shrub (to 1.3 m tall), as it grows in areas that are infrequently traveled. It differs from the other members of this genus in having tiny, scalelike leaves.

160. *Euphorbia milii* (crown-of-thorns), flowers

161. *Euphorbia tirucalli* (pencil tree) and *Crinum latifolium* (crinum lily) with flowers

Scientific Name: Hibiscus rosa-sinensis L. (Photo 162)
Common Names: Chinese hibiscus, rose-of-China, *peregrina, cucarda*

Family: Malvaceae (Mallow)

Range: Cultivated escape; also known from other tropical regions throughout the world, probably originally from tropical Asia, perhaps China.

Islands Inhabited: Floreana, San Cristóbal, Santa Cruz

Habitat: Arid lowlands and moist uplands

Description: Shrub or small tree typically to 4 m tall, occasionally taller. *Leaves* alternate, simple, ovate to broadly ovate, 4–15 cm long, surfaces smooth and glossy, margins serrate. *Flowers* axillary, solitary; corolla red, sometimes darker at the base, petals 5, 5–9 cm long; stamens numerous, united in a tubular column, filaments red, anthers yellow with pollen; style branches red. *Fruit* a capsule, ovoid, to ca. 2 cm long, with 5 sections (rarely 10); seeds numerous.

Comments: Hibiscus rosa-sinensis is a popular ornamental, and its beautiful flowers are commonly observed in and around each town. Occasionally one may encounter specimens with double flowers, or flowers of a different color, such as pink or orange. This is due to the fact that many different cultivars of this species exist, and they are capable of hybridizing with one another. The specific epithet of this species, *rosa-sinensis,* means "rose-of-China."

162. *Hibiscus rosa-sinensis* (Chinese hibiscus), flower

Scientific Name: Hibiscus schizopetalus (Dyer) Hook. f. (Photo 163)
Common Names: Chinese lantern, coral hibiscus, *farolillo chino*

Family: Malvaceae (Mallow)

Range: Cultivated; also known from other tropical regions throughout the world, originally from tropical eastern Africa.

Islands Inhabited: Santa Cruz

Habitat: Arid lowlands and moist uplands

Description: Shrub or small tree to ca. 3 m tall, with many slender, drooping branches. *Leaves* alternate, simple, ovate, to ca. 10 cm long, margins toothed. *Flowers* axillary, solitary; corolla red with whitish or pink streaks, petals 5, to ca. 6 cm long, dissected and curved backward; stamens numerous, united in a tubular column, filaments red; style branches red. *Fruit* a capsule, with 5 sections (rarely 10); seeds numerous.

Comments: Hibiscus diversifolius Jacq. (spiny hibiscus), another member of the genus, occurs on Isabela (SN), San Cristóbal, and Santa Cruz. This introduced species differs from the others primarily in having stems that are typically spiny. Its petals (4–6 cm long) are whitish, yellow, or reddish purple, with maroon at the base.

163. *Hibiscus schizopetalus* (Chinese lantern), flower

Scientific Name: Rubus niveus Thunb. (Photo 164)
Common Name: Hill raspberry

Family: Rosaceae (Rose)

Range: Cultivated escape; also known from other tropical regions throughout the world, perhaps originally from India.

Islands Inhabited: San Cristóbal, Santa Cruz

Habitat: Moist uplands

Description: Shrub, stems to ca. 2 m long, covered with hooked prickles to ca. 7 mm long. *Leaves* alternate, odd-pinnately compound, with prickles; leaflets 5–9, elliptic-ovate, to ca. 5 cm long, lower surface covered with white hairs, margins serrate. *Flowers* in terminal panicles with prickles, corolla pinkish purple, 4–5 mm long, petals 5; stamens numerous. *Fruit* an aggregate composed of numerous fleshy drupelets, dark red to reddish black, roundish, ca. 1 cm in diameter; seeds 1 per drupelet.

Comments: This species, originally planted on San Cristóbal, has now escaped cultivation and become a nuisance in the highlands. It remains to be seen whether this plant will become as much of a pest as *Cinchona succirubra* (quinine tree), *Psidium guajava* (common guava), and *Lantana camara* (multicolored lantana).

Rubus bogotensis Kunth is also cultivated. It is known only from Volcán Sierra Negra (Isabela).

164. *Rubus niveus* (hill raspberry), flowers and fruits

Scientific Name: Castela galapageia Hook. f. (Photo 165)
Common Names: Castela, *amargo*

Family: Simaroubaceae (Simarouba)

Range: Endemic

Islands Inhabited: Baltra, Española, Fernandina, Floreana, Isabela (A,D,W), Marchena, Pinta, Pinzón, Rábida, San Cristóbal, Santa Cruz, Santa Fe, Santiago, Seymour, Islet(s)

Habitat: Arid lowlands

Description: Shrub to 5 m tall, much-branched, stems usually possessing axillary spines 1.5–7 mm long, young stems covered with fuzzy, white hairs. *Leaves* alternate or in clusters of 2–5, simple, variable in shape, typically oblong to elliptic, 8–40 mm long, lower surface densely covered with white hairs, margins entire, wavy, or slightly toothed. *Flowers* unisexual (plants dioecious), in axillary clusters or occasionally solitary; corolla red (outside) and pale yellow (inside) when mature, ca. 4 mm across, petals 4, 3–5 mm long; staminate flowers with 8 stamens. *Fruit* composed of 1–4 drupes, bright red, 8–12 mm long, fleshy; seed 1 per drupe.

Comments: This shrub typically grows on dry, rocky lava areas. It has been suggested that goats refrain from eating this plant due to its bitter taste.

165. *Castela galapageia* (castela), flower and fruits

See also

Bauhinia monandra (p. 72)
Capraria biflora (p. 116)
Encelia hispida (p. 124)
Piscidia carthagenensis (p. 55)

Shrubs with Alternate Leaves and Green Flowers

Scientific Name: Maytenus octogona (L'Hér.) DC. (Photos 166–67)
Common Names: Maytenus, *arrayancillo, rompe ollas*

Family: Celastraceae (Staff-tree)

Range: Native; also known from western South America (Ecuador to Chile).

Islands Inhabited: Baltra, Española, Fernandina, Floreana, Isabela (A,CA,D,SN,W), Pinzón, Rábida, San Cristóbal, Santa Cruz, Santa Fe, Santiago, Seymour, Islet(s)

Habitat: Coastal zone and arid lowlands

Description: Shrub or small tree to 8 m tall. *Leaves* alternate, simple, usually elliptic or obovate, ca. 2–5 cm long, somewhat leathery, margins entire, wavy or dentate. *Flowers* usually perfect (occasionally unisexual if flowers abort one or the other sex, then plants monoecious or dioecious), solitary or in short axillary panicles; corolla greenish, 3–4 mm across, petals 5; stamens 5. *Fruit* a capsule, green, somewhat roundish and 3-angled, 9–12 mm long, 8–10 mm across, opening into 3 sections; seeds 1–3 (usually 1), covered by a red, fleshy aril.

Comments: This plant is extremely common on several of the islands. Medium ground finches (*Geospiza fortis*) find the taste of the colorful aril irresistible.

166. (above) *Maytenus octogona* (maytenus), flowers

167. (left) *Maytenus octogona* (maytenus), fruits

Scientific Name: Atriplex peruviana Moq. (Photo 168)
Common Names: Atriplex

Family: Chenopodiaceae (Goosefoot)

Range: Native; also known from Peru and Chile.

Islands Inhabited: Española, Pinzón, Santa Cruz, Seymour, Wolf

Habitat: Coastal zone and arid lowlands

Description: Shrub to 1 m tall, much-branched. *Leaves* alternate, simple, blade ovate, ovate-rhombic, or rounded, to 4 cm long, both surfaces covered with scalelike material, margins entire. *Flowers* unisexual (plants monoecious or dioecious). Staminate flowers clustered in terminal or axillary spikes or panicles; calyx greenish, sepals 5; corolla absent; stamens 5. Pistillate flowers axillary or in clusters at the bottom of staminate inflorescences; each flower with two small bracts beneath; calyx absent; corolla absent. *Fruit* a utricle, enclosed by two bracts that enlarge as it matures, bracts typically 4–6 mm long and 3–4 mm wide; seed 1.

Comments: This plant is an important component of the vegetation at Punta Suarez (Española). Some members of this genus are used on the mainland as salad greens, while others are used as forage.

168. *Atriplex peruviana* (atriplex), flowers

Scientific Name: Ricinus communis L. (Photo 169)
Common Names: Castor bean, *higuerilla*

Family: Euphorbiaceae (Spurge)

Range: Cultivated escape; also known from other tropical regions throughout the world, originally from Africa.

Islands Inhabited: Baltra, Floreana, Isabela (SN), San Cristóbal, Santa Cruz

Habitat: Arid lowlands

Description: Shrub or small tree to 5 m tall, containing a watery sap. *Leaves* alternate, simple, to 1 m long and ca. 1 m wide, usually palmately 7- to 9-lobed, lobe margins serrate. *Flowers* unisexual or bisexual (plants monoecious), in terminal panicles. Staminate flowers on lower half of panicle; calyx greenish, 5-lobed, lobes 4–9 mm long; corolla absent; stamens numerous. Pistillate flowers on upper half of panicle; calyx typically greenish, 5-lobed, 3–4 mm long; corolla absent; style lobes bright red. Bisexual flowers, when present, on upper half of panicle. *Fruit* a capsule, roundish, 1–2 cm in diameter, covered with dark brown soft spines; seeds 3–6, mottled.

Comments: The genus name, *Ricinus*, comes from the Latin word meaning "tick" because the plump, mottled seed resembles a tick in appearance. Like other members of the spurge family, this is a poisonous plant. Eating any part, especially the seed coat, can cause death. Symptoms include a burning sensation in the mouth and throat, nausea, and convulsions. Simply touching the leaves may cause itching and sneezing in some individuals. However, this plant has also provided many useful substances. For example, its seeds are the source of castor oil. Not only is this oil of benefit medicinally, but it has also been used in the production of soap, lubricants, margarine, and paints. Castor bean grows in just about any soil with good drainage, especially in disturbed areas such as roadsides. This native of tropical Africa is sometimes cultivated for its beautiful leaves.

169. *Ricinus communis* (castor bean), flowers

Scientific Name: Urera caracasana (Jacq.) Griseb. (Photo 170)
Common Names: Tree nettle, *ortiga de arbol*

Family: Urticaceae (Nettle)

Range: Native; also known from other regions of tropical America.

Islands Inhabited: Isabela (SN), Santa Cruz, Santiago

Habitat: Moist uplands

Description: Shrub or small tree to 10 m tall, all new growth with stinging hairs. *Leaves* alternate, simple, broadly ovate to roundish, typically 15–20 cm long, lower surface hairy, margins crenate-dentate. *Flowers* unisexual (plants dioecious), in axillary cymes. Staminate flowers greenish, 2–2.5 mm across, perianth segments 4; stamens 4–5. Pistillate flowers greenish, perianth segments 4, ca. 3 mm long, the inner pair slightly longer than the outer, and almost twice as wide. Inner pair becomes fleshy and bright orange in fruit. *Fruit* an achene, ca. 1–1.5 mm long; achene plus the surrounding fleshy tissue ca. 3–4 mm across.

Comments: Tree nettle typically inhabits areas with abundant shade. The genus name comes from the Latin *uro*, "to burn or sting," and refers to the plant's stinging hairs.

170. *Urera caracasana* (tree nettle), flowers and fruits

See also

Conocarpus erectus (p. 78)
Croton scouleri (p. 53)
Cryptocarpus pyriformis (p. 115)

Shrubs with Alternate Leaves and Brown Flowers

Scientific Name: Tournefortia psilostachya Kunth (Photos 171–72)
Common Name: Smooth-stemmed tournefortia, *palito negro*

Family: Boraginaceae (Borage)

Range: Native; also known from western South America (Colombia to Peru, and extending partially into Brazil).

Islands Inhabited: Española, Floreana, Isabela (CA,D,SN,W), Pinta, Pinzón, San Cristóbal, Santa Cruz, Santiago, Islet(s)

Habitat: Arid lowlands

Description: Vinelike perennial shrub, woody at the base, spreading over other vegetation and rocks, stems smooth. *Leaves* alternate, simple; blade ovate, elliptic, or ovate-lanceolate, 1–4 cm wide, both surfaces covered with fine white hairs, margins entire. *Flowers* in axillary and terminal, often branching, scorpioid cymes, corolla tube greenish yellow to brownish, with 5 reddish brown lobes, tube 4–5 mm long, lobes 2–2.5 mm long; stamens 5. *Fruit* a drupe consisting of 2–4 nutlets, yellow to yellowish orange, often with a few black spots, roundish, 5–6 mm in diameter; seed 1 per nutlet.

Comments: This plant's common name, *palito negro,* means "black stick" and refers to the older branches, which take on a distinctive dark brown to black color. If one looks closely, tiny white dots known as "lenticels" may be seen covering these branches. *Tournefortia* was named for the French botanist Joseph Pitton de Tournefort (1656–1708), who established the concept of genus.

171. (above) *Tournefortia psilostachya* (smooth-stemmed tournefortia), flowers

172. (left) *Tournefortia psilostachya* (smooth-stemmed tournefortia), flowers and fruits

Shrubs with Opposite Leaves and White Flowers

Scientific Name: Alternanthera echinocephala (Hook. f.) Christoph. (Photo 173)

Common Name: Spiny-headed chaff flower

Family: Amaranthaceae (Amaranth)

Range: Native; also known from Peru.

Islands Inhabited: Española, Floreana, Isabela (A,CA,E,SN,W), Pinta, Pinzón, San Cristóbal, Santa Cruz, Santa Fe, Santiago, Islet(s)

Habitat: Arid lowlands

Description: Shrub to 2 m tall, numerous upright branches. *Leaves* opposite, simple, lanceolate, 4–9 cm long, margins entire. *Flowers* in terminal and axillary headlike spikes 1–1.5 cm across, each flower with 1 bract and 2 bracteoles beneath, these white to greenish white, sometimes tinged with pink; calyx white to greenish white, sepals 5, to 10 mm long; corolla absent; stamens 5. *Fruit* a utricle, 2 mm long; seed 1.

Comments: This is one of 13 species of *Alternanthera* occurring in the Galápagos, and due to its relatively large, spiny heads, it is the easiest to identify. Exposed pollen on the anthers gives the flowers their touch of yellow. Six of the other 12 species in this genus are endemics.

173. *Alternanthera echinocephala* (spiny-headed chaff flower), flowers

Scientific Name: Alternanthera filifolia (Hook. f.) J. T. Howell
(Photos 174–75)
Common Name: Thread-leafed chaff flower

Family: Amaranthaceae (Amaranth)

Range: Endemic

Islands Inhabited: Española, Fernandina, Floreana, Isabela (A,D,SN), Pinta, Pinzón, Rábida, San Cristóbal, Santa Cruz, Santa Fe, Santiago, Islet(s)

Habitat: Arid lowlands

Description: Shrub to 1.5 m tall, numerous upright branches. *Leaves* opposite, simple, linear to oblong-oblanceolate, 2–5 cm long, 2–7 mm wide, margins entire. *Flowers* in terminal or axillary headlike spikes 3–15 mm long and 3–5 mm across, solitary or in clusters of 2–7, each flower with 1 bract and 2 bracteoles beneath, these white to greenish white; calyx white to greenish white, sepals 5, to 3 mm long; corolla absent; stamens 5. *Fruit* a utricle; seed 1.

Comments: Alternanthera filifolia includes seven subspecies. The difficult task of identifying these is simplified by the facts that subsp. *glauca* J. T. Howell is found only on the islet of Tortuga near Isabela; subsp. *pintensis* Eliasson is known only from Pinta; and subsp. *rabidensis* Eliasson is restricted to Rábida. Each of these is considered rare, and no other subspecies share their islands. *Alternanthera filifolia* subsp. *filifolia* (Photos 174–75) is the only subspecies that occurs on Española, Fernandina, Isabela (A,D,SN), Pinzón, Santa Cruz, and Santa Fe. Identification becomes a challenge when more than one subspecies inhabits a single island. For example, both subsp. *filifolia* and subsp. *glaucescens* (Hook. f.) Eliasson occur on San Cristóbal. However, the former has stems and leaves that are not glaucous and leaves that are not fleshy, while the latter has stems and leaves that are glaucous and leaves that are somewhat fleshy. *Alternanthera filifolia* subsp. *filifolia* shares Santiago with subsp. *microcephala* Eliasson (also rare), but these two differ in that the former has rounded or oblong, headlike spikes 7–15 mm long, while the latter has rounded, headlike spikes that are 3–4 mm across. Both subsp. *filifolia* and subsp. *nudicaulis* (Hook. f.) Eliasson inhabit Floreana. The former has flower tips that are usually straight, while the latter has flower tips that are often incurved. It should be noted that not all taxonomists agree with this classification. Some feel that the above-mentioned differences do not merit separation into subspecies. Additional studies on this species will be necessary for more definitive conclusions.

174. *Alternanthera filifolia* subsp. *filifolia* (thread-leafed chaff flower)

175. *Alternanthera filifolia* subsp. *filifolia* (thread-leafed chaff flower), flowers

Scientific Name: Alternanthera halimifolia (Lam.) Standl. (Photo 176)
Common Names: Chaff flower, *monte colorado*

Family: Amaranthaceae (Amaranth)

Range: Native; also known from other regions of tropical America (Mexico to Chile).

Islands Inhabited: Fernandina, Isabela (A,CA,SN,W), Pinta, San Cristóbal, Santa Cruz, Santiago

Habitat: Arid lowlands and moist uplands

Description: Shrub to 1 m tall, numerous stems. *Leaves* opposite, simple, elliptic, 2–10 cm long, both surfaces with short stellate hairs, margins entire. *Flowers* in axillary headlike spikes 6 mm across, each flower with 1 bract and 2 bracteoles beneath, these white to greenish white; calyx white to greenish white, sepals 5, to 4 mm long; corolla absent; stamens 5. *Fruit* a utricle; seed 1.

Comments: This shrub is classified as rare. The specific epithet comes from the Greek *halimos*, "of the sea." Mature pollen released from the anthers gives the flowers their touch of yellow.

176. *Alternanthera halimifolia* (chaff flower), flowers

Scientific Name: Scalesia crockeri J. T. Howell (Photos 177–78)
Common Names: Crocker's scalesia, *Lechoso*

Family: Asteraceae (Sunflower)

Range: Endemic

Islands Inhabited: Baltra, Santa Cruz

Habitat: Arid lowlands

Description: Shrub typically less than 1 m tall. *Leaves* opposite or almost opposite, simple, ca. 4–8 cm long (including petiole); petiole winged; blade ovate to broadly elliptic or almost roundish, both surfaces glandular and rough, margins serrate or dentate to entire. *Flowers* in discoid heads, disc to 2 cm across, flowers white, typically 20–60 per head, perfect. *Fruit* an achene, flattened, to 5 mm long; seed 1.

Comments: Scalesia crockeri is considered rare. It is the only member of the genus that occurs on Baltra.

177. (right) *Scalesia crockeri* (Crocker's scalesia), flowers

178. (below) *Scalesia crockeri* (Crocker's scalesia)

Scientific Name: Scalesia gordilloi O. Ham. & Wium-Anders.

(Photos 179–80; see also Photo 14)

Common Names: Gordillo's scalesia, *lechoso*

Family: Asteraceae (Sunflower)

Range: Endemic

Islands Inhabited: San Cristóbal

Habitat: Arid lowlands

Description: Shrub to 1.5 m tall. *Leaves* opposite or almost opposite, simple; blade ovate, to ca. 6 cm long, both surfaces hairy and somewhat glandular, margins entire to occasionally minutely serrate. *Flowers* in discoid heads, disc to 1.2 cm across, flowers white, ca. 30–50 per head, perfect. *Fruit* an achene, flattened, to 5 mm long, with a few small bumps or awns on top, to 1.5 mm long; seed 1.

Comments: This plant was named in honor of the Galápagos naturalist Jacinto Gordillo. Hamann and Wium-Andersen (1986) described the species after it was brought to their attention by Gordillo, who was working on San Cristóbal at the time as a representative of the CDRS. Its status is vulnerable.

179. (below) *Scalesia gordilloi* (Gordillo's scalesia)

180. (right) *Scalesia gordilloi* (Gordillo's scalesia), flowers

Scientific Name: Scalesia helleri B. L. Rob. (Photo 181)
Common Names: Heller's scalesia, *lechoso*

Family: Asteraceae (Sunflower)

Range: Endemic

Islands Inhabited: Santa Cruz, Santa Fe

Habitat: Arid lowlands

Description: Shrub, typically 1–2.5 m tall, occasionally somewhat treelike. *Leaves* opposite or alternate, simple; blade elliptic, ovate, triangular, or irregular, ca. 4–10 cm long, lobed, sinuses extending almost to the midrib, each lobe divided again, both surfaces glandular and sometimes hairy. *Flowers* generally in discoid heads, disc to 1.7 cm across; disc flowers white, ca. 30–100 per head, perfect; occasionally a few irregular ray flowers also present, white, sterile. *Fruit* an achene, flattened, 3–4 mm long, occasionally with a few small bumps on top; seed 1.

Comments: This species includes two subspecies. *Scalesia helleri* subsp. *helleri* possesses outer bracts beneath each head that are oblong-ovate or oblong, while the outer bracts of subsp. *santacruziana* Harling (Photo 181) are oblong-obovate or obovate. *Scalesia helleri* subsp. *helleri* is known from Santa Cruz and Santa Fe and is considered rare, whereas subsp. *santacruziana* occurs only on Santa Cruz and is considered endangered. Both subspecies are typically found growing on relatively inaccessible cliffs, especially near the coast. It appears that this distribution is the result of grazing pressure caused by feral goats over the years. A thorough search near Tortuga Bay (Santa Cruz) will yield a few specimens of subsp. *santacruziana.* The leaves of this plant, when rubbed between the fingers, produce a pleasant odor.

181. *Scalesia helleri* subsp. *santacruziana* (Heller's scalesia), flowers

Scientific Name: Chamaesyce amplexicaulis (Hook. f.) Burch (Photo 182)
Common Name: Chamaesyce

Family: Euphorbiaceae (Spurge)

Range: Endemic

Islands Inhabited: Baltra, Daphne, Floreana, Genovesa, Marchena, Rábida, San Cristóbal, Santa Cruz, Santiago, Wolf, Islet(s)

Habitat: Arid lowlands

Description: Small shrub to 60 cm tall, containing a milky sap, stems swollen at the nodes. *Leaves* opposite, simple; blade roundish, 3–11 mm long, 4–11 mm wide, cordate base wrapping around the stem, margins entire. *Flowers* unisexual (plants monoecious), in terminal (but appearing axillary), solitary, greenish cup-shaped structures known as "cyathia." Each cyathium has 4 black glands that possess white to yellowish green appendages resembling petals. Staminate flowers 8–14 per cyathium; calyx absent; corolla absent; stamen 1. Pistillate flowers 1 per cyathium, projecting from the mouth of the cyathium, greenish; calyx absent; corolla absent. *Fruit* a capsule, broadly ovoid, to 1.8 mm long, 2.3 mm across, smooth; seed 1.

Comments: This is one of 11 species of *Chamaesyce* that occurs in the archipelago. This particular species inhabits both sandy soils and lava. The actual flowers are so tiny that they are often overlooked. However, the overall appearance of this plant makes it difficult to misidentify.

Of the 11 species inhabiting the archipelago, 8 are endemics and 3 have been introduced. The endemic *Chamaesyce viminea* (Hook. f.) Burch (spurred chamaesyce) (Photo 183) is one of the more easily identified members of the genus. This much-branched shrub may reach a height of 1.5 m. Its leaves are opposite or clustered on short spur shoots, and the leaf blades are linear (ca. 1–4 cm long). Its flowers are similar to those of *C. amplexicaulis*, but there are no petal-like appendages. This plant typically inhabits the arid lowlands of Fernandina, Floreana, Genovesa, Isabela (A,D), Marchena, Pinta, Rábida, San Cristóbal, Santa Cruz, Santa Fe, Santiago, Seymour, and one or more of the smaller islands. During the cool season the whole plant takes on a characteristic reddish brown color.

182. *Chamaesyce amplexicaulis* (chamaesyce), flowers

183. *Chamaesyce viminea* (spurred chamaesyce)

Scientific Name: Chiococca alba (L.) Hitchc. (Photo 184)
Common Names: Milkberry, *espuela de gallo*

Family: Rubiaceae (Madder)

Range: Native; also known from other regions of tropical America (Mexico and Florida to northern South America).

Islands Inhabited: Española, Fernandina, Floreana, Isabela (A,CA,D,SN), Marchena, Pinta, Pinzón, San Cristóbal, Santa Cruz, Santiago

Habitat: Arid lowlands and moist uplands

Description: Shrub or small tree 2–6 m tall. *Leaves* opposite, simple; blade ovate, lanceolate, or elliptic, 2–10 cm long, somewhat leathery, upper surface dark, shiny green, margins entire. *Flowers* in axillary racemes or panicles; corolla white to yellowish white, quickly turning yellow with age, funnelform with 5 lobes, 6–8 mm long; stamens 5. *Fruit* a drupe, white, roundish and flattened, 4–6 mm long, leathery; seeds 2.

Comments: The genus name comes from the Greek *chion*, "snow," and *kokkos*, "berry," an obvious reference to the white fruits produced by this plant. Milkberry is a common element of forested areas in the archipelago.

184. *Chiococca alba* (milkberry), flowers and fruits

Scientific Name: Coffea arabica L. (Photos 185–86)
Common Names: Coffee, *cafeto, café*

Family: Rubiaceae (Madder)

Range: Cultivated escape; also known from other tropical regions throughout the world, originally from northeastern Africa, probably Ethiopia.

Islands Inhabited: Floreana, Isabela (SN), San Cristóbal, Santa Cruz

Habitat: Moist uplands

Description: Shrub or small tree to 8 m tall. *Leaves* opposite, simple; blade lance-elliptic, 8–20 cm long, somewhat leathery, upper surface dark, shiny green, margins entire. *Flowers* in axillary clusters; corolla white, salverform with 5 lobes, 1.5–2 cm long; stamens 5. *Fruit* berrylike, red to dark red or black, ellipsoid to roundish, ca. 1–1.5 cm long, fleshy; seeds 2, 9–11 mm long, grooved on flat side.

Comments: This plant's "beans" are well known as the source of the beverage coffee. However, it is actually the seeds that are used for this purpose. This plant has now escaped into some of the forested areas on the inhabited islands. It is easily recognizable by its attractive white flowers, which also have a pleasant fragrance.

185. *Coffea arabica* (coffee), flowers

186. *Coffea arabica* (coffee), fruits

Scientific Name: Psychotria rufipes Hook. f. (Photo 187)
Common Names: White wild coffee, *cafetillo*

Family: Rubiaceae (Madder)

Range: Endemic

Islands Inhabited: Fernandina, Isabela (A,CA), Pinta, San Cristóbal, Santa Cruz, Santiago

Habitat: Arid lowlands and moist uplands

Description: Shrub 1–3 m tall, stems somewhat hairy. *Leaves* opposite, simple; blade broadly elliptic-ovate to obovate, 8–20 cm long, leathery, dark green, lower surface with reddish hairs, especially at the angles where lateral veins join the midrib, margins entire. *Flowers* in terminal and axillary panicles; corolla white, tubular-salverform with 5 lobes, 8–10 mm across, tube 6–7 mm long, lobes 3–4 mm long, throat densely covered with hairs; stamens 5. *Fruit* berrylike, red, ellipsoid to roundish, 7–9 mm long, fleshy; seeds 2, longitudinal ridges on outer surface.

Comments: The common name, "white wild coffee," comes from this plant's resemblance to *Coffea arabica.* However, *Coffea* has flowers and fruits that are noticeably larger.

Psychotria angustata Andersson (pink wild coffee), an endangered endemic, is known only from Floreana. It differs from *P. rufipes* primarily in having stems and leaves that are smooth, corollas that are pink to rose-colored, and leaves that are shorter (8–10 cm) and broadly lanceolate.

187. *Psychotria rufipes* (white wild coffee), fruits

Scientific Name: Clerodendrum molle HBK. (Photos 188–89)
Common Names: Glorybower, *rodilla de caballo*

Family: Verbenaceae (Vervain)

Range: One variety is endemic, the other native.

Islands Inhabited: Floreana, Isabela (CA,SN), San Cristóbal, Santa Cruz, Santiago

Habitat: Arid lowlands

Description: Shrub to 5 m tall, much-branched, stems densely covered with short hairs. *Leaves* opposite or in whorls of 3, simple; blade elliptic, 1–6 cm long, margins entire. A sharpish protuberance is left on the stem after each leaf drops. *Flowers* in axillary cymes; corolla pinkish purple (tube) and white (lobes) with a pinkish purple throat, salverform with 5 lobes, 2.5–3 cm long; stamens 4. *Fruit* drupaceous, grayish brown, roundish, ca. 8 mm in diameter, often separating into 4 sections when mature; seed 1 per section.

Comments: This species includes two varieties. *Clerodendrum molle* var. *molle* (Photos 188–89) possesses leaves that are densely covered with short hairs underneath. It is found throughout the lowlands on Floreana, Isabela (CA,SN), San Cristóbal, Santa Cruz, and Santiago. This native is known from western South America (Panama to Peru) as well.

Clerodendrum molle var. *glabrescens* Svens., considered endemic and rare, is known only from Santa Cruz. It differs from var. *molle* in having leaves that are relatively smooth underneath. It should be noted that some taxonomists do not consider this character significant enough to warrant formal taxonomic recognition.

The flowers of glorybower have a pleasing fragrance and are a source of nectar for the Galápagos carpenter bee (*Xylocopa darwini*). However, this bee normally removes the nectar in what might be considered a roundabout way. Since it isn't able to reach the nectar from the front of the flower, it cuts a small slit in the base of the flower tube. From this location, it can easily feed on the nectar. This behavior is known as "nectar robbing." Hawk moths also visit these plant for nectar, usually after dusk.

Another member of this genus, *C. philippinum* Schauer (Philippine clerodendrum) (Photo 190), has recently been introduced to the islands. It is presently known only as an ornamental in and around the town of El Progreso (San Cristóbal) but should be monitored so that it does not spread to surrounding areas. Its attractive flowers are rather unusual in that the reproductive parts have been modified into petals. Consequently, this plant does not produce fruits (Wagner et al. 1990). Instead, it spreads entirely by means of root suckers.

188. *Clerodendrum molle* var. *molle* (glorybower), flowers

189. (above) *Clerodendrum molle* var. *molle* (glorybower), flowers

190. (right) *Clerodendrum philippinum* (Philippine clerodendrum), flowers

Scientific Name: Lantana peduncularis Andersson (Photo 191)
Common Names: Galápagos lantana, *supi-rosa*

Family: Verbenaceae (Vervain)

Range: Endemic

Islands Inhabited: Española, Fernandina, Floreana, Genovesa, Isabela (A,D,SN), Marchena, Pinta, Pinzón, Rábida, San Cristóbal, Santa Cruz, Santa Fe, Santiago, Islet(s)

Habitat: Arid lowlands

Description: Shrub to 2 m tall, stems covered with hairs. *Leaves* opposite, simple; blade ovate or ovate-lanceolate to elliptic, to 7 cm long, to 4 cm wide, both surfaces hairy, sometimes velvety, margins serrate to somewhat crenate. *Flowers* in axillary heads; corolla white, often with a yellow throat, salverform with 5 unequal lobes, 8–9 mm long; stamens 4. *Fruit* drupaceous, roundish, often separating into 2 sections when mature; seed 1 per section.

Comments: Lantana peduncularis often forms dense thickets on lava and cinder fields. These thickets are difficult to negotiate, especially during the cool season, when the plants are stiff and dry. Wiggins and Porter (1971) list two varieties in this species. They suggest that *L. peduncularis* var. *macrophylla* Moldenke differs from var. *peduncularis* primarily in having larger leaves and a persistent, dense velvety covering on both leaf surfaces. However, these differences are not consistent, and all individuals are treated here as members of a single variable species.

191. *Lantana peduncularis* (Galápagos lantana), flowers

Scientific Name: Lippia rosmarinifolia Andersson (Photo 192)
Common Name: Narrow-leafed lippia

Family: Verbenaceae (Vervain)

Range: Endemic

Islands Inhabited: Fernandina, Isabela (A,D,SN), Pinta, Santiago

Habitat: Arid lowlands and moist uplands

Description: Shrub ca. 2 m tall, occasionally taller, much-branched, stems covered with short hairs. *Leaves* opposite, simple; blade linear to narrowly elliptic, ca. 3 cm long and 7 mm wide, upper surface rough, lower surface covered with stiff hairs, margins entire, toothed, or pinnately lobed, and rolled under. *Flowers* in axillary heads, each flower with a green bracteole beneath; corolla white to pale purplish, with a yellow throat, salverform with 4 lobes, somewhat 2-lipped; stamens 4. *Fruit* drupaceous, ovoid, ca. 1 mm long, separating into 2 sections when mature; seed 1 per section.

Comments: Lippia rosmarinifolia frequents lava and cinder fields at various altitudes but is considered rare. Wiggins and Porter (1971) mention three varieties within the species, var. *rosmarinifolia*, var. *latifolia* Moldenke, and var. *stewartii* Moldenke. However, only the first two are currently accepted (van der Werff 1977).

Lippia salicifolia Andersson (willow-leafed lippia), another endemic, resembles *L. rosmarinifolia* in being a shrub. It differs in having oblong-lanceolate leaf blades up to 7.5 cm long and 2.5 cm wide. This species is considered endangered and is known only from Floreana. The genus name honors the seventeenth-century Italian naturalist Auguste Lippi.

192. *Lippia rosmarinifolia* (narrow-leafed lippia), flowers

See also

Avicennia germinans (p. 80)
Galvezia leucantha (p. 166)
Laguncularia racemosa (p. 81)
Psidium galapageium (p. 83)
Rhizophora mangle (p. 86)
Scalesia divisa (p. 105)

Shrubs with Opposite Leaves and Yellow or Orange Flowers

Scientific Name: Lecocarpus darwinii Adsersen (Photos 193–94)
Common Name: Curve-spined lecocarpus

Family: Asteraceae (Sunflower)

Range: Endemic

Islands Inhabited: San Cristóbal

Habitat: Arid lowlands

Description: Shrub to 1.5 m tall, branches numerous but with a single stem at the base. *Leaves* opposite, simple; petiole winged; blade lanceolate to elliptic, 3–10 cm long, margins rolled under, serrate on upper portion, entire on lower portion. *Flowers* in radiate heads; heads solitary, borne near the branch tips; disc flowers yellow, staminate; ray flowers yellow, pistillate. *Fruit* an achene, 5–8 mm long, with one or more curved spines of variable length; seed 1.

Comments: Lecocarpus is one of seven angiosperm genera restricted to the Galápagos. This species is the least likely to be seen by tourists, as it does not occur at the more popular visitor sites. It is considered endangered.

193. (left) *Lecocarpus darwinii* (curve-spined lecocarpus), flowers

194. (above) *Lecocarpus darwinii* (curve-spined lecocarpus), flowers and fruits

Scientific Name: Lecocarpus lecocarpoides (B. L. Rob. & Greenm.) Cronq. & Stuessy (Photos 195–96)
Common Name: Straight-spined lecocarpus

Family: Asteraceae (Sunflower)

Range: Endemic

Islands Inhabited: Española, Islet(s)

Habitat: Arid lowlands

Description: Shrub to 1 m tall, branches numerous but with a single stem at the base. *Leaves* opposite, simple; petiole without wings or only slightly winged; blade ovate to elliptic, typically 1–6 cm long, deeply pinnately dissected, with each lobe secondarily divided. *Flowers* in radiate heads; heads solitary, borne near the branch tips; disc flowers yellow, staminate; ray flowers yellow, pistillate. *Fruit* an achene, 4–5 mm long, with one or more straight spines, each approximately the same length as the body of the fruit or shorter; seed 1.

Comments: According to Adsersen (1980), *Lecocarpus lecocarpoides* also includes those plants listed in Wiggins and Porter (1971) as *L. leptolobus* (Blake) Cronq. & Stuessy. *Lecocarpus lecocarpoides* is easily differentiated from *L. darwinii* (curve-spined lecocarpus) by its dissected leaves. In addition, it usually has fewer ray flowers per head (5–8) than *L. darwinii* (8–11). *Lecocarpus lecocarpoides* is considered endangered.

195. (above) *Lecocarpus lecocarpoides* (straight-spined lecocarpus), flowers

196. (left) *Lecocarpus lecocarpoides* (straight-spined lecocarpus), flowers and fruits

Scientific Name: Lecocarpus pinnatifidus Decne. (Photos 197–98)
Common Name: Wing-fruited lecocarpus

Family: Asteraceae (Sunflower)

Range: Endemic

Islands Inhabited: Floreana

Habitat: Arid lowlands

Description: Shrub to 2 m tall, branches numerous but with a single stem at the base. *Leaves* opposite, simple; petiole without wings; blade elliptic, typically 3–7 cm long, deeply pinnately to thrice-pinnately dissected into oblong segments. *Flowers* in radiate heads; heads solitary, borne near the branch tips; disc flowers yellow, staminate; ray flowers yellow, pistillate. *Fruit* an achene, 2–4 mm long, with a broad wavy-edged wing or collar around the top, to 1.5 cm across; seed 1.

Comments: The specific epithet of this plant, *pinnatifidus*, refers to the numerous divisions of each leaf blade. This species is considered vulnerable.

197. *Lecocarpus pinnatifidus* (wing-fruited lecocarpus), flowers

198. *Lecocarpus pinnatifidus* (wing-fruited lecocarpus), flowers and fruits

Scientific Name: Macraea laricifolia Hook. f. (Photos 199–200)
Common Names: Macraea, *romerillo*

Family: Asteraceae (Sunflower)

Range: Endemic

Islands Inhabited: Fernandina, Floreana, Isabela (A,D,SN), Pinta, Rábida, San Cristóbal, Santa Cruz, Santiago

Habitat: Arid lowlands and moist uplands

Description: Shrub to 2.5 m tall, numerous slender branches. *Leaves* opposite or in small clusters, simple, linear and very narrow, 1.5–5 cm long, margins rolled under. *Flowers* generally in radiate heads, disc 6–12 mm across; heads borne near the branch tips; disc flowers yellow, perfect; ray flowers, when present, light yellow, pistillate. *Fruit* an achene, 2–4.5 mm long, winged, somewhat rough in texture, topped by a crown of small teethlike projections; seed 1.

Comments: This species belongs to one of only seven angiosperm genera endemic to the Galápagos. It is named for James Macrae, who, during a visit in 1825, collected plants on Isabela. The specific epithet alludes to the clusters of narrow leaves like those of the conifer genus *Larix*, the common larch.

199. (below) *Macraea laricifolia* (macraea), flowers

200. (right) *Macraea laricifolia* (macraea), flowers

Scientific Name: Lantana camara L. (Photo 201)
Common Name: Multicolored lantana

Family: Verbenaceae (Vervain)

Range: Cultivated escape; also known from other tropical regions throughout the world, probably originally from the West Indies.

Islands Inhabited: Floreana, San Cristóbal, Santa Cruz

Habitat: Arid lowlands and moist uplands

Description: Shrub to 3 m tall, stems covered with hairs, usually armed with prickles. *Leaves* opposite, simple; blade ovate to oblong, ca. 2–12 cm long, upper surface rough, lower surface somewhat hairy, margins serrate. *Flowers* in axillary heads; corolla opening yellowish orange to orange, usually turning pink, red, or reddish orange, rarely white, salverform with 5 unequal lobes, ca. 1 cm long; stamens 4. *Fruit* drupaceous, dark purple or black, roundish, ca. 3–4 mm in diameter, often separating into 2 sections when mature; seed 1 per section.

Comments: Multicolored lantana has escaped from cultivation and become a serious problem on Floreana. Methods for its control are currently being studied. All parts of this plant, especially the leaves and green berries, are toxic to humans and livestock. Handling the leaves also causes itching in some people.

201. *Lantana camara* (multicolored lantana), flowers and fruits

See also

Avicennia germinans (p. 80)
Chiococca alba (p. 152)
Rhizophora mangle (p. 86)

Shrubs with Opposite Leaves and Pink, Red, or Purple Flowers

Scientific Name: Miconia robinsoniana Cogn. (Photos 202–3; see also Photo 31)
Common Names: Galápagos miconia, *cacaotillo*

Family: Melastomataceae (Melastome)

Range: Endemic

Islands Inhabited: San Cristóbal, Santa Cruz

Habitat: Moist uplands

Description: Shrub 2–5 m tall, much-branched, young branches 4-angled. *Leaves* opposite, simple, oblanceolate to elliptic or oblong, ca. 10–30 cm long, leathery, usually with 3 prominent veins, margins entire. *Flowers* in terminal panicles to 18 cm long, corolla purplish, petals 4 or 5, 6–7 mm long; stamens 8–10, anthers curved. *Fruit* a berry, dark purplish, roundish, 5–6 mm high; seeds numerous.

Comments: The status of Galápagos miconia is vulnerable, and its area of coverage is not nearly as extensive as in the past. On Santa Cruz, much of the *Miconia* zone has been replaced by exotics, especially *Cinchona succirubra* (quinine tree).

202. (above) *Miconia robinsoniana* (Galápagos miconia), flowers and fruits

203. (left) *Miconia robinsoniana* (Galápagos miconia), flowers and fruits

Scientific Name: Galvezia leucantha Wiggins (Photo 204)
Common Name: Galápagos shrub snapdragon

Family: Scrophulariaceae (Figwort)

Range: Endemic

Islands Inhabited: Fernandina, Isabela (A), Rábida, Santiago

Habitat: Arid lowlands

Description: Shrub to 1.5 m tall, much-branched, stems smooth to densely glandular-hairy. *Leaves* opposite, simple; blade elliptic-lanceolate to ovate-lanceolate, 1.5–3.5 cm long, smooth to densely glandular-hairy, margins entire. *Flowers* axillary, solitary; corolla white to white and reddish or pinkish, tubular, 2-lipped, upper lip 2-lobed, lower lip 3-lobed, tube 5–6.5 mm long; stamens 5, 1 nonfunctional. *Fruit* a capsule, roundish, 3–4.5 mm long; seeds numerous.

Comments: Three subspecies of Galápagos shrub snapdragon have been suggested by Elisens (1989). *Galvezia leucantha* subsp. *leucantha*, considered endemic and rare, is found on Fernandina and Isabela (A). Its stems and leaves are usually smooth or only slightly hairy, the corolla is completely white, and the corolla tube is 5.5–6.5 mm long.

A second subspecies, as yet unnamed, is known only from Santiago. It also has stems and leaves that are typically smooth or slightly hairy and corolla tubes that are 5.5–6.5 mm long. However, it differs from the above subspecies in having a corolla that is white and reddish or pinkish.

Galvezia leucantha subsp. *pubescens* Wiggins is considered in danger of extinction (Lawesson 1990), but the plant can still be found on Rábida. Its stems and leaves are densely glandular-hairy, the corolla is white or rarely white tinged with red, and the corolla tube is 5–5.5 mm long.

The plant pictured here, which was photographed on Rábida, is perhaps a fourth subspecies. It possesses stems and leaves that are densely glandular-hairy and corollas that are conspicuously reddish purple on the outside and pink to white on the inside. Its corolla tubes have never been measured. Future studies should determine whether this is indeed a separate subspecies or simply a more colorful representative of subsp. *pubescens.*

Elisens (1989) has noted that *G. leucantha* is pollinated by the Galápagos carpenter bee (*Xylocopa darwini*) but that its closest mainland relative and probable ancestor, *G. fruticosa* Gmelin., is pollinated by hummingbirds and has red flowers with long corolla tubes (9–14 mm). His explanation for this is that hummingbirds do not inhabit the archipelago, and therefore the Galápagos shrub snapdragon's flowers have become adapted for pollination by the carpenter bee. This adaptation was necessary because the carpenter bee's mouthparts are not suitable for collecting nectar at the bottom of a long corolla tube, and bees are rarely attracted to completely red flowers.

204. *Galvezia leucantha* (Galápagos shrub snapdragon), flowers

Scientific Name: Stachytarpheta cayennensis (L. C. Rich.) M. Vahl

(Photo 205)

Common Names: False vervain, brazilian tea

Family: Verbenaceae (Vervain)

Range: Introduced; also known from other tropical regions throughout the world, originally ranging from Mexico and Cuba to Peru and Argentina.

Islands Inhabited: Floreana, Isabela (SN), San Cristóbal, Santa Cruz

Habitat: Arid lowlands

Description: Shrub 1–2.5 m tall, much-branched. *Leaves* opposite, simple; blade ovate to elliptic, 3–7 cm long, upper surface somewhat rough, margins serrate. *Flowers* in terminal spikes to 34 cm long, each flower with a bracteole beneath; corolla purplish and white to bluish and white, salverform with 5 lobes, ca. 5 mm across; stamens 2. *Fruit* a schizocarp, enclosed in calyx, oblong, separating into 2 sections when mature; seed 1 per section.

Comments: The genus name of this shrub is derived from the Greek *stachys*, "spike," and *tarphys*, "dense," in reference to the crowded inflorescence.

205. *Stachytarpheta cayennensis* (false vervain), flowers

See also

Lantana camara (p. 164)
Lippia rosmarinifolia (p. 159)
Psychotria angustata (p. 155)

Shrubs with Opposite Leaves and Blue Flowers

See

Stachytarpheta cayennensis (p. 167)

Shrubs with Opposite Leaves and Green Flowers

Scientific Name: Froelichia juncea B. L. Rob. & Greenm. (Photos 206–7)
Common Name: Froelichia

Family: Amaranthaceae (Amaranth)

Range: Endemic

Islands Inhabited: Isabela (A,D,SN), Santa Cruz

Habitat: Arid lowlands

Description: Shrub to 1 m tall, branching. *Leaves* opposite, simple, linear, 2–8 cm long except for upper ones, which are much reduced (ca. 2 mm long), margins entire. *Flowers* in terminal spikes 1.5–5 cm long, each flower with 1 bract and 2 bracteoles beneath, these greenish; calyx greenish, 3.5–4.5 mm long, 5-lobed; corolla absent; stamens 5. *Fruit* a utricle; seed 1.

Comments: This species includes two subspecies, both considered rare. *Froelichia juncea* subsp. *alata* J.T. Howell (Photos 206–7), found only on Santa Cruz, forms a calyx-tube with two thin wings around each fruit. Each of the wings is ca. 1 mm wide, and the calyx-tube is often covered with minute, woolly hairs. *Froelichia juncea* subsp. *juncea*, known from Isabela (A,D,SN), differs in that it typically does not form wings on the calyx-tube. However, if present, they are thick and only ca. 0.5 mm wide. Additionally, the calyx-tube is not covered with woolly hairs.

Froelichia nudicaulis Hook. f., also an endemic, is somewhat shorter (to 0.5 m) and bushier than *F. juncea* and has spikes that are blunt at the tip rather than pointed. This species is divided into three subspecies. *Froelichia nudicaulis* subsp. *curta* J. T. Howell is found only on Pinzón and is considered rare. It has rounded spikes, woolly calyx-tubes, and wings that are ca. 1 mm wide. *Froelichia nudicaulis* subsp. *nudicaulis*, also rare, occurs on Floreana, San Cristóbal, and Santiago. It too has woolly calyx-tubes, and wings that are ca. 1 mm wide, but it differs in having oblong spikes. Finally, subsp. *lanigera* (Andersson) Eliasson occurs on Fernandina and Isabela (A,D,W). It produces both rounded and oblong spikes and has woolly calyx-tubes. However, it differs in that its calyx-tubes are typically wingless.

206. (above) *Froelichia juncea* subsp. *alata* (froelichia)

207. (left) *Froelichia juncea* subsp. *alata* (froelichia), flowers

Scientific Name: Batis maritima L. (Photos 208–9)
Common Name: Saltwort

Family: Bataceae (Batis)

Range: Native; also known from other tropical regions throughout the world (southern United States to northern South America).

Islands Inhabited: Baltra, Floreana, Isabela (SN,W), San Cristóbal, Santa Cruz, Santiago, Islet(s)

Habitat: Coastal zone

Description: Shrub to 1.5 m tall, stems fleshy. *Leaves* opposite, simple, oblanceolate, 1–2 cm long, fleshy, margins entire. *Flowers* unisexual (plants dioecious), in axillary conelike spikes, each flower with a bract beneath. Staminate spikes 3–8 mm long, greenish; calyx cup-shaped, greenish; corolla absent; stamens 4, staminodia 4. Pistillate spikes 5–20 mm long, greenish; calyx absent; corolla absent; stigma cushionlike in appearance. *Fruit* drupelike, those of each spike fused together into a common structure; seeds 4.

Comments: The specific epithet of saltwort, *maritima*, refers to its coastal habitat. In fact, a nice population can be seen near the beach at Tortuga Bay (Santa Cruz). If one looks carefully, the minute flowers may be observed, although they are often overlooked.

208. *Batis maritima* (saltwort)

209. *Batis maritima* (saltwort), flowers and fruits

Scientific Name: Phoradendron henslowii (Hook. f.) B. L. Rob. (Photo 210)
Common Names: Galápagos mistletoe, foradendron, *suelda con suelda*

Family: Viscaceae (Mistletoe)

Range: Endemic

Islands Inhabited: Fernandina, Floreana, Isabela (A,CA,D,SN), Pinta, Pinzón, Rábida, San Cristóbal, Santa Cruz, Santiago

Habitat: Arid lowlands and moist uplands

Description: Shrub to ca. 1 m tall, parasitic on woody plants but possessing green, photosynthetic leaves. *Leaves* opposite, simple, mainly lance-ovate to elliptical, to 20 cm long, somewhat leathery, margins entire. *Flowers* unisexual (plants monoecious), somewhat embedded in axillary spikes, staminate and pistillate flowers randomly arranged on each inflorescence; tepals yellowish green, ca. 2 mm across; staminate flowers with 3 stamens. *Fruit* a drupe, white and pinkish, spherical, 4–6 mm in diameter, with a watery pulp; seed 1.

Comments: Common hosts of this plant include *Croton scouleri* (Galápagos croton), *Macraea laricifolia* (macraea), and *Zanthoxylum fagara* (cat's claw). Individuals growing in the arid lowlands are typically smaller than those inhabiting moister areas (Stewart 1911). The fruits of the Galápagos mistletoe differ from those of most members of the genus in that they are not sticky.

210. *Phoradendron henslowii* (Galápagos mistletoe), flowers and fruits

See also

Chamaesyce amplexicaulis (p. 151)

Shrubs with Whorled Leaves and White Flowers

See

Clerodendrum molle (p. 156)
Nerium oleander (p. 175)

Shrubs with Whorled Leaves and Pink, Red, or Purple Flowers

Scientific Name: Nerium oleander L. (Photo 211)
Common Name: Common oleander

Family: Apocynaceae (Dogbane)

Range: Cultivated; also known from other tropical regions throughout the world, originally from the Mediterranean region.

Islands Inhabited: Santa Cruz

Habitat: Arid lowlands

Description: Shrub to 6 m tall. *Leaves* in whorls of 3, simple, linear to oblong-lanceolate, to 25 cm long, margins entire. *Flowers* in terminal cymes; corolla pink or white, funnelform with 5 lobes, to 5 cm across, 5 hairy scales in the throat; stamens 5. *Fruit* a follicle, 10–15 cm long; seeds numerous.

Comments: This attractive shrub is commonly used as a landscape plant around the town of Puerto Ayora. However, all of its parts are extremely poisonous to humans and livestock. Ingesting one leaf may prove fatal, and some even believe that eating food cooked over the wood of this plant may cause death. Symptoms include nausea, vomiting, and decreased heartbeat. A double-flowered cultivar of this species is occasionally encountered.

211. *Nerium oleander* (common oleander), flowers

Scientific Name: Russelia equisetiformis Schlecht. & Cham. (Photo 212)
Common Name: Coral plant

Family: Scrophulariaceae (Figwort)

Range: Cultivated; also known from other tropical regions throughout the world, probably originally from Mexico.

Islands Inhabited: Santa Cruz

Habitat: Arid lowlands

Description: Shrub to ca. 1.5 m tall, much-branched, drooping, secondary stems arranged in whorls. *Leaves* in whorls of 3–6, simple, ovate or elliptic, 1–2 cm long, short-lived, margins toothed, mostly reduced to scales. *Flowers* in axillary clusters of 1–3; corolla red, tubular, ca. 2.5 cm long, somewhat 2-lipped, upper lip 2-lobed, lower lip 3-lobed; stamens 4. *Fruit* a capsule, roundish; seeds numerous.

Comments: The genus *Russelia* was named in honor of Alexander Russell (1715–68), a Scottish physician and naturalist. Coral plant may be seen growing as an ornamental in Puerto Ayora.

212. *Russelia equisetiformis* (coral plant), flowers

Scientific Name: Lycium minimum C. L. Hitchc. (Photo 213)
Common Name: Galápagos lycium

Family: Solanaceae (Potato)

Range: Endemic

Islands Inhabited: Baltra, Española, Floreana, Isabela (SN), Pinta, Pinzón, Rábida, Santa Cruz, Santa Fe, Islet(s)

Habitat: Coastal zone

Description: Shrub 1–3 m tall, major branches with shorter, spine-tipped lateral branches. *Leaves* in clusters of 2–5, simple, narrowly elliptic to somewhat club-shaped, 4–12 mm long, fleshy, margins entire. *Flowers* axillary, solitary or in small clusters; corolla white with purplish veins, tubular with 4 lobes, tube 2–3.5 mm long, lobes 1.5–2 mm long; stamens 4. *Fruit* a berry, reddish orange, obovoid, 5–6 mm long; seeds 2.

Comments: The spiny lateral branches of this rare shrub make it easily identifiable, even during the cool season, when all of its leaves are absent.

213. *Lycium minimum* (Galápagos lycium)

Scientific Name: Nolana galapagensis (Christoph.) I. M. Johnst. (Photos 214–15)
Common Name: Galápagos clubleaf

Family: Nolanaceae (Nolana)

Range: Endemic

Islands Inhabited: Floreana, Isabela (A,W), San Cristóbal, Santa Cruz, Seymour, Islet(s)

Habitat: Coastal zone

Description: Shrub to 1.5 m tall, with numerous, short, leafy, lateral branches. *Leaves* in crowded clusters, simple, club-shaped, to 3.5 cm long, fleshy, margins entire. *Flowers* axillary, solitary; corolla white, funnelform with 5 lobes, ca. 1–2 cm long; stamens 5. *Fruit* composed of 15–25 sections; seeds few to numerous per section.

Comments: The genus name comes from the Latin *nola*, "small bell," in reference to the flower shape. Galápagos clubleaf is considered rare, but it is present at a few of the more popular visitor sites.

214. (above) *Nolana galapagensis* (Galápagos clubleaf) and red-footed boobies (*Sula sula*)

215. (right) *Nolana galapagensis* (Galápagos clubleaf), flower

See also

Bursera graveolens (p. 48)
B. malacophylla (p. 50)
Carica papaya (p. 51)
Grabowskia boerhaaviaefolia (p. 120)

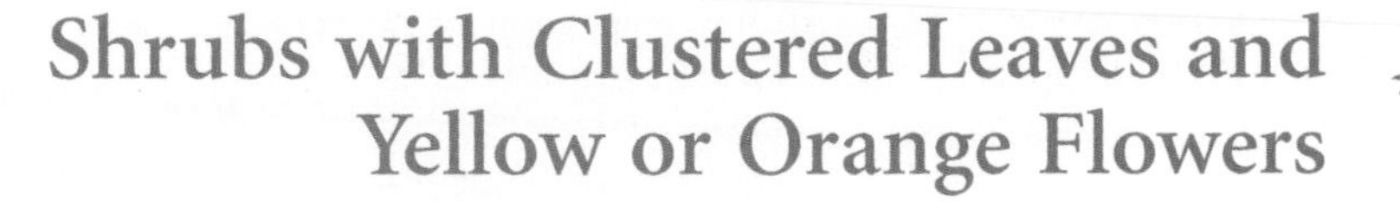

Shrubs with Clustered Leaves and Yellow or Orange Flowers

See

Castela galapageia (p. 138)
Macraea laricifolia (p. 163)

Shrubs with Clustered Leaves and Pink, Red, or Purple Flowers

See

Castela galapageia (p. 138)

Shrubs with Clustered Leaves and Green Flowers

See

Chamaesyce viminea (p. 151)

Herbs with Alternate Leaves and White Flowers

Scientific Name: Heliotropium angiospermum Murr. (Photo 216)
Common Names: Heliotrope, *cola de escorpion*

Family: Boraginaceae (Borage)

Range: Native; also known from other regions of tropical America (southern Florida to Argentina).

Islands Inhabited: Daphne, Española, Fernandina, Floreana, Genovesa, Isabela (A,CA,D, SN,W), Pinta, Pinzón, Rábida, San Cristóbal, Santa Cruz, Santa Fe, Santiago, Seymour, Wolf, Islet(s)

Habitat: Arid lowlands

Description: Low-growing perennial to 50 cm tall, lower parts somewhat woody, upper parts herbaceous. *Leaves* alternate, simple, elliptical, ovate, or sometimes lanceolate, 1.5–10 cm long, 1–2 cm wide, with white hairs, margins entire. *Flowers* in scorpioid spikes to 20 cm long; corolla white, often with a yellowish green throat, campanulate with 5 lobes, 1.5–2 mm long, throat hairy; stamens 5. *Fruit* consisting of 2 nutlets, 2–2.8 mm across; seeds 1–2 per nutlet.

Comments: It's possible that one might confuse this plant with *Heliotropium anderssonii* B. L. Rob. (Andersson's heliotrope), a vulnerable endemic known only from Santa Cruz. However, the latter has narrower leaves (2–7 mm wide) with light brown hairs, and pale yellow flowers.

Heliotropium rufipilum (Benth.) I. M. Johnst. var. *anademum* I. M. Johnst. is an introduced species found only on Floreana. It differs from *H. angiospermum* in that its flowers are 3–4 mm long.

216. *Heliotropium angiospermum* (heliotrope), flowers and fruits

Scientific Name: Heliotropium curassavicum L. (Photo 217)
Common Names: Seaside heliotrope, *cola de escorpion*

Family: Boraginaceae (Borage)

Range: Native; also known from other tropical regions throughout the world.

Islands Inhabited: Española, Floreana, Genovesa, Isabela (A,SN), Marchena, Pinta, Pinzón, Rábida, San Cristóbal, Santa Cruz, Santiago, Seymour, Islet(s)

Habitat: Coastal zone

Description: Low-growing perennial herb, stems somewhat fleshy, smooth, and with a bluish, waxy covering, stem tips curving upward. *Leaves* alternate, simple, elliptic or oblanceolate, 1.5–6 cm long, somewhat fleshy, margins entire. *Flowers* in scorpioid spikes 10–15 cm long, often 2-cleft; corolla white with a yellowish throat that may turn purplish, funnelform with 5 lobes, ca. 2 mm long; stamens 5. *Fruit* consisting of 4 nutlets, 2–2.5 mm across; seed 1 per nutlet.

Comments: All Galápagos members of this species represent var. *curassavicum.* Seaside heliotrope is an example of a halophyte, which means that it has adapted to living in soils with high concentrations of salt. Consequently, it is often found near alkaline flats and in coastal areas. Its smooth, somewhat fleshy leaves distinguish it from the other heliotropes found in the archipelago.

217. *Heliotropium curassavicum* var. *curassavicum* (seaside heliotrope), flowers and fruits

Scientific Name: Tiquilia galapagoa (J. T. Howell) A. Richardson (Photos 218–19)
Common Name: Gray matplant

Family: Boraginaceae (Borage)

Range: Endemic

Islands Inhabited: Baltra, Daphne, Fernandina, Floreana, Genovesa, Isabela (D,SN), Marchena, Pinta, Rábida, San Cristóbal, Santa Cruz, Santiago, Islet(s)

Habitat: Arid lowlands

Description: Low-growing perennial herb to 50 cm tall, woody near the base, typically spreading and matlike, sometimes upright, grayish white, hairy. *Leaves* alternate, in closely crowded clusters, simple, variously shaped but primarily ovate, 2–6 mm long, hairy, margins rolled under. *Flowers* borne in dense clusters, only 1 or a few flowers open per cluster at any one time; corolla white, consisting of a campanulate to funnelform or almost cylindrical tube with 5 lobes, 3–4 mm long; stamens 5. *Fruit* consisting of 4 nutlets, dull black, less than 1 mm long, surface granular in appearance; seed 1 per nutlet.

Comments: Tiquilia, as a genus, is quite easy to identify, especially during the archipelago's cool season, when the older growth takes on a light gray color. However, the three species are somewhat difficult to distinguish without careful study. *Tiquilia galapagoa,* which is defined here to include both *Coldenia fusca* (Hook. f.) A. Gray and *C. galapagoa* Howell in Wiggins and Porter (1971), differs from *T. darwinii* and *T. nesiotica* primarily in having nutlets with a dull black, granular surface.

218. (above) *Tiquilia galapagoa* (gray matplant)

219. (right) *Tiquilia galapagoa* (gray matplant), flowers

Scientific Name: Tiquilia nesiotica (J. T. Howell) A. Richardson

(Photos 220–21; see also Photo 23)

Common Name: Gray matplant

Family: Boraginaceae (Borage)

Range: Endemic

Islands Inhabited: Santiago, Islet(s)

Habitat: Arid lowlands

Description: Low-growing perennial herb to 40 cm tall, woody near the base, usually spreading and matlike, sometimes upright, grayish white, hairy. *Leaves* alternate, in closely crowded clusters, simple, ovate to ovate-lanceolate, 1.5–2.5 mm long, hairy, margins rolled under. *Flowers* borne in dense clusters, only 1 or a few flowers open per cluster at any one time; corolla white, consisting of a cylindrical tube with 5 lobes, 2.5–3 mm long; stamens 5. *Fruit* consisting of 4 nutlets, shiny black, less than 1 mm long, surface mostly smooth in appearance; seed 1 per nutlet.

Comments: Considered rare, this spectacular gray-colored plant thrives in the sand and volcanic ash of Bartolomé and Santiago. Though it appears lifeless during the cool season, it is actually very much alive and will become active again once the rains begin. At this time, new green leaves may be seen among the older gray-colored growth. This plant is referred to as *Coldenia nesiotica* Howell in Wiggins and Porter (1971).

Tiquilia darwinii (Hook. f.) A. Richardson (gray matplant), which Wiggins and Porter (1971) list as *C. darwinii* (Hook. f.) A. Gray, differs primarily in having a slightly shorter corolla (1.5–2.5 mm) with a campanulate to funnelform tube. In addition, it normally does not grow as tall as *T. nesiotica*. It occurs on Baltra, Española, Isabela (D,SN), Santa Cruz, Santa Fe, Santiago, and one or more of the smaller islands.

220. (above) *Tiquilia nesiotica* (gray matplant)

221. (left) *Tiquilia nesiotica* (gray matplant), flowers

Scientific Name: Evolvulus simplex Andersson (Photo 222)
Common Name: White evolvulus

Family: Convolvulaceae (Morning-glory)

Range: Native; also known from mainland Ecuador and Peru.

Islands Inhabited: Floreana, Genovesa, Isabela (A,D), Rábida, San Cristóbal, Santa Cruz, Santiago

Habitat: Arid lowlands

Description: Erect annual herb to 40 cm tall, stems simple or occasionally branched, covered with hairs. *Leaves* alternate, simple; blade lanceolate to oblong-lanceolate, 10–22 mm long, both surfaces covered with hairs, margins entire. *Flowers* axillary, solitary or in clusters of 2–3; corolla white, somewhat funnelform-rotate, 3–4 mm long, lobes ca. 0.5 mm long; stamens 5. *Fruit* a capsule, somewhat spherical, ca. 2 mm across; seeds 1–4, pale brown.

Comments: Both this species and *E. convolvuloides* (purple evolvulus) can be seen on Rábida.

222. *Evolvulus simplex* (white evolvulus), flowers

Scientific Name: Anoda acerifolia DC. (Photo 223)
Common Name: Anoda

Family: Malvaceae (Mallow)

Range: Introduced; also known from other tropical regions throughout the world, originally from tropical America.

Islands Inhabited: Floreana, Santa Cruz

Habitat: Moist uplands

Description: Erect or sprawling annual herb to ca. 70 cm tall, stems covered with bristles. *Leaves* alternate, simple to palmately 5-lobed, somewhat triangular, to ca. 5 cm long, margins entire or serrate near the base. *Flowers* axillary, solitary; corolla white, petals 5, 1.2–1.5 cm long; stamens numerous, united in a tubular column. *Fruit* a schizocarp, disc-shaped, ca. 1–1.5 cm across, with 9–15 sections, beaked, hairy; seed 1 per section, brown.

Comments: This attractive herb is common along the roadsides in the highlands of Santa Cruz. Its specific epithet refers to the leaves, which resemble those found in the maple genus, *Acer*.

223. *Anoda acerifolia* (anoda), flower and fruits

Scientific Name: Desmanthus virgatus (L.) Willd. (Photo 224)
Common Name: Slender mimosa

Family: Mimosaceae (Mimosa)

Range: Native; also known from other tropical regions throughout the world, originally from tropical America.

Islands Inhabited: Española, Floreana, Pinta, Pinzón, San Cristóbal, Santa Cruz, Islet(s)

Habitat: Arid lowlands and moist uplands

Description: Perennial herb with a woody base or a shrub to ca. 2 m tall. *Leaves* alternate, bipinnately compound with 1–7 pairs of pinnae; leaflets 10–20 pairs per pinna, oblong, 2–8 mm long, margins occasionally with fine hairs. *Flowers* in axillary clusters, greenish white; petals 5, 2.5–4 mm long; stamens 10. *Fruit* a legume, linear, ca. 1.5–5.5 cm long, flattened; seeds 10–25, brown.

Comments: All Galápagos members of this species represent var. *depressus* (Willd.) B. L. Turner. According to Wagner et al. (1990), it is used in Hawaii as forage for livestock.

224. *Desmanthus virgatus* var. *depressus* (slender mimosa), flowers and fruits

Scientific Name: Musa x *paradisiaca* L. (Photos 225–26)
Common Names: Banana, plantain, *banano, plátano*

Family: Musaceae (Banana)

Range: Cultivated escape; also known from other tropical regions throughout the world.

Islands Inhabited: Santa Cruz

Habitat: Moist uplands

Description: Treelike perennial herb to 8 m tall. *Leaves* alternate but clustered near the top, simple; blade 2–2.5 m long, margins entire at first but often forming tears that extend to the midrib. *Flowers* functionally unisexual (plants monoecious), in large, terminal, pendent spikes with pistillate flowers in clusters toward the base and staminate flowers in clusters toward the tip, each flattened cluster with a purplish bract beneath; corolla yellowish white, petal 1; stamens 5. *Fruit* a berry, greenish, cylindrical and strongly angled, 15–30 cm long, with cream-colored flesh; seeds abortive.

Comments: Musa acuminata Colla is similar to this species but is shorter (to 6 m tall), has shorter leaf blades (ca. 2 m long), and produces smaller fruits (8–12 cm long) with an orange to reddish flesh. It is cultivated on Floreana and Santa Cruz. *Musa* x *paradisiaca* originated in southeast Asia as a hybrid between *M. acuminata* and *M. balbisiana* Colla. Cultivars of both *M.* x *paradisiaca* and *M. acuminata* are grown for their edible fruits. The fruits of the plantain or *plátano* are edible when cooked. They are prepared in a variety of ways, such as fried *patacones,* and often served as a side dish. The fruits of the banana plant, or *banano,* are bright yellow when ripe. They are soft, sweet, and edible without cooking. In the islands this fruit goes by the name *guineo.*

225. (above) *Musa x paradisiaca* (plantain), fruits

226. (left) *Musa x paradisiaca* (plantain), flowers

Scientific Name: Habenaria monorrhiza (Sw.) Rchb. f. (Photo 227)
Common Name: Fringed orchid

Family: Orchidaceae (Orchid)

Range: Native; also known from other regions of tropical America (Central America and the West Indies to Peru and Brazil).

Islands Inhabited: Isabela (SN), Santa Cruz

Habitat: Moist uplands

Description: Erect perennial herb to 1.2 m tall, stem often dark-spotted. *Leaves* alternate, simple, ovate to oblong-lanceolate, to 15 cm long, sheathing at base, margins entire. *Flowers* in terminal racemes to 25 cm long, each flower with a bract beneath; corolla white, petals 3; stamen 1. The two lateral petals are up to 8 mm long. The third petal, referred to as the "lip" or "labellum," is 3-lobed at one end and somewhat fringed. The other end forms a nectar-containing spur to 24 mm long. *Fruit* a capsule, brown, ellipsoid; seeds numerous, dustlike.

Comments: Habenaria alata Hook., also native, differs from *H. monorrhiza* in that its stem is not dark-spotted, its flowers are greenish to greenish yellow, and the labellum is unlobed. It is found on Fernandina and Isabela (A,SN).

Habenaria distans Griseb., another native, also has greenish flowers, but the labellum is 3-lobed. It is found only on Isabela (A).

227. *Habenaria monorrhiza* (fringed orchid), flowers and fruits

Scientific Name: Ionopsis utricularioides (Sw.) Lindl. (Photo 228)
Common Name: Ionopsis

Family: Orchidaceae (Orchid)

Range: Native; also known from other regions of tropical America (Florida and the West Indies to Paraguay).

Islands Inhabited: Floreana, Isabela (A,SN), Pinta, Pinzón, Rábida, Santa Cruz, Santiago

Habitat: Arid lowlands and moist uplands

Description: Perennial herb, epiphytic, growing on trees. *Leaves* alternate, simple, narrowly oblong or strap-shaped, to 16 cm long, leathery or fleshy, sheathing at base, margins entire. *Flowers* in terminal racemes or panicles to 60 cm long, each flower with a small bract beneath; corolla white, pinkish, or white with reddish purple stripes, petals 3; stamen 1. The two lateral petals are ovate-oblong, to 6 mm long. The third petal, referred to as the "lip" or "labellum," is basally attached to a central structure known as the "column" or "gynandrium," which is a compound structure composed of the stamen, style, and stigma. The labellum itself is roundish, deeply 2-lobed, and to 2 cm long. *Fruit* a capsule, brown, ca. 2 cm long; seeds numerous, dustlike.

Comments: The genus name of this orchid is derived from the Greek *ion*, "violet," and *opsis*, "resemblance." This is because the flowers superficially resemble those in the genus *Viola*.

228. *Ionopsis utricularioides* (ionopsis), flowers

Scientific Name: Plumbago scandens L. (Photos 229–30)
Common Name: White leadwort

Family: Plumbaginaceae (Plumbago)

Range: Native; also known from other regions of tropical America (Arizona and Florida to Argentina).

Islands Inhabited: Española, Fernandina, Floreana, Isabela (A,CA,D,SN), Pinta, Pinzón, San Cristóbal, Santa Cruz, Santiago, Islet(s)

Habitat: Arid lowlands and moist uplands

Description: Perennial herb, sprawling on the ground or climbing over rocks and plants, stems 1–6 m long, often dark red. *Leaves* alternate, simple, elliptic, oblong, or ovate, to 10 cm long, often dark red, margins entire. *Flowers* in terminal spikes; calyx covered with sticky, glandular hairs, often dark red; corolla white or tinged with pink, salverform with 5 lobes, tube 2–2.5 cm long, lobes 6–10 mm long; stamens 5, bluish. *Fruit* a capsule, somewhat ovoid, 5–7 mm long; seed 1.

Comments: This species commonly grows in forested areas, as well as in disturbed sites such as roadside ditches. Its flowers are often visited by the Galápagos sulfur butterfly (*Phoebis sennae*) and the Galápagos blue butterfly (*Leptotes parrhasioides*). Galápagos doves (*Zenaida galapagoensis*) on Pinta are known to eat the fruits of this plant.

Plumbago coerulea HBK. (blue leadwort), also native, is found only on Volcán Wolf (Isabela). It differs primarily in having smaller flowers (tube 8–10 mm long) that are purplish blue in color. In addition, the lower half of the calyx is smooth rather than covered with glandular hairs.

229. (above) *Plumbago scandens* (white leadwort), flowers and fruits

230. (right) *Plumbago scandens* (white leadwort), flowers and fruits

Scientific Name: Polygonum galapagense Caruel (Photo 231)
Common Name: Galápagos knotweed

Family: Polygonaceae (Buckwheat)

Range: Endemic

Islands Inhabited: Isabela (SN), San Cristóbal, Santa Cruz

Habitat: Moist uplands

Description: Perennial herb to 1.2 m tall, stems smooth, with distinctly swollen joints. *Leaves* alternate, simple; blade broadly ovate to lance-elliptic, 8–14 cm long, 2–4.5 cm wide, both surfaces covered with golden yellow hairs, margins entire. The basal portion of each leaf forms a membraneous sheath (ocrea) that surrounds the stem. This ocrea is 1.5–2.5 cm long and has golden yellow bristles 5–12 mm long on its upper margin. *Flowers* in axillary racemes 5–9 cm long; calyx white, sepals 5; corolla absent; stamens 3–9. *Fruit* an achene, dark brown to black, shiny, lens-shaped, ca. 3 mm long and 3 mm wide; seed 1.

Comments: The genus derives its name from the Greek *poly*, "many," and *gonu*, "knee." This refers to the plant's jointed stems. Galápagos knotweed normally inhabits moist areas in the highlands and is considered rare.

A similar species is the native *Polygonum acuminatum* HBK. from San Cristóbal. It differs primarily in having longer ocreae (2–4 cm).

231. *Polygonum galapagense* (Galápagos knotweed), flowers

Scientific Name: Polygonum opelousanum Small (Photo 232)
Common Name: Knotweed

Family: Polygonaceae (Buckwheat)

Range: Native; also known from other regions of tropical America (Mexico and southeastern U.S. to northern South America).

Islands Inhabited: Floreana, Isabela (A,SN), San Cristóbal, Santa Cruz

Habitat: Moist uplands

Description: Perennial herb to 1 m tall, stems smooth, with distinctly swollen joints. *Leaves* alternate, simple; blade narrowly lanceolate, 3–12 cm long, 5–15 mm wide, lower surface with scattered, grayish glandular dots, margins entire. The basal portion of each leaf forms a membraneous sheath (ocrea) that surrounds the stem. This ocrea is 1–1.5 cm long and has yellowish bristles 5 mm long on its upper margin. Its surface is covered with sharp, yellow, upwardly turned hairs. *Flowers* in axillary racemes 1.5–4 cm long; calyx white to pinkish, 1.5–2 mm long, sepals 5; corolla absent; stamens 7–8. *Fruit* an achene, dark reddish brown, shiny, 3-sided, 1.5–2 mm long and 1.2–2 mm wide; seed 1.

Comments: This species normally inhabits areas in the highlands such as wet meadows. A similar species, the native *Polygonum hydropiperoides* Michx. var. *persicarioides* (HBK.) Stanford (mild water-pepper), has pinkish flowers tinged with green. It differs from *P. opelousanum* primarily in having longer achenes (2.5–3 mm). In addition, it does not have grayish glandular dots on the lower surface of its leaves. It is known only from Isabela (SN).

Polygonum punctatum Ell. (water knotweed), a native found on San Cristóbal, possesses flowers that are primarily greenish. It differs from all other members of the genus in having sepals with conspicuous glandular dots.

232. *Polygonum opelousanum* (knotweed), flowers

Scientific Name: Calandrinia galápagosa H. St. John (Photo 233)
Common Names: Galápagos rock-purslane

Family: Portulacaceae (Purslane)

Range: Endemic

Islands Inhabited: San Cristóbal

Habitat: Arid lowlands

Description: Perennial herb to 60 cm tall, slightly woody at the base, stems fleshy. *Leaves* alternate, simple, linear, 3–7 cm long, somewhat fleshy, margins entire. *Flowers* in terminal cymes; tepals 5, white, ca. 1 cm long; stamens 12–15. *Fruit* a capsule, yellowish, roundish, ca. 4 mm in diameter; seeds numerous, dark reddish brown.

Comments: At one time this species was known to frequent lava fissures at Sappho Cove on the northwest side of San Cristóbal. However, it may no longer exist at that location. In fact, it is considered endangered. An attempt is currently being made by the CDRS to save this species by germinating seeds in cultivation and returning the seedlings to their natural habitat.

233. *Calandrinia galapagosa* (Galápagos rock-purslane), flower

Scientific Name: Capsicum frutescens L. (Photo 234)
Common Names: Bird pepper, tabasco pepper

Family: Solanaceae (Potato)

Range: Cultivated escape; also known from other tropical regions throughout the world, originally from tropical America.

Islands Inhabited: Floreana, Isabela (A,SN), Santa Cruz, Santiago

Habitat: Arid lowlands

Description: Perennial herb or small shrub to 2 m tall, much-branched. *Leaves* alternate, simple; blade ovate to elliptic-ovate, 2.5–6 cm long, margins entire. *Flowers* typically in axillary clusters of 2–4, occasionally solitary; corolla greenish white to greenish yellow, somewhat rotate with 5 lobes, 8–10 mm across; stamens 5, anthers bluish. *Fruit* a berry, red, oblong-ovoid, 8–15 mm long; seeds numerous, yellowish.

Comments: The genus is thought to have derived its name from the Greek *kapto*, meaning "to bite." This refers to the hot, biting flavor of the fruit. However, one does not have to taste the fruit to feel its sting. Simply touching it or the seeds, then inadvertently rubbing one's eyes, will result in a burning sensation not soon to be forgotten.

Two cultivated members of the genus are found in the archipelago. These are *Capsicum annuum* L. (cayenne pepper) on Santa Cruz and *C. pendulum* Willd. on San Cristóbal. The fruits of these plants can be used to make the popular condiment ají, which is seen on virtually every dining table in the archipelago.

Capsicum galapagoense Hunz. (Galápagos pepper) is the only endemic member of the genus. It has white flowers with yellow or tan markings in the throat and yellow anthers. Known only from Isabela (A,SN) and Santa Cruz, it is considered rare.

234. *Capsicum frutescens* (bird pepper), flowers and fruits

Scientific Name: Exedeconus miersii (Hook. f.) D'Arcy (Photo 235)
Common Name: Galápagos shore petunia

Family: Solanaceae (Potato)

Range: Endemic

Islands Inhabited: Daphne, Española, Fernandina, Floreana, Genovesa, Isabela (A,CA,D, E,SN,W), Marchena, Pinta, Pinzón, Rábida, San Cristóbal, Santa Cruz, Santiago, Islet(s)

Habitat: Coastal zone

Description: Low-growing annual herb, stems trailing over the ground, covered with glandular hairs. *Leaves* alternate or almost opposite, simple; blade ovate to ovate-reniform, 4–20 cm long, both surfaces hairy, margins irregularly few-toothed, sometimes wavy. *Flowers* axillary, solitary, but sometimes appearing clustered; corolla white with yellowish green veins, funnelform with 5 teeth or lobes, 3–7 cm long, lobes 2–10 mm long; stamens 5. *Fruit* a capsule, elongate-ovate; seeds numerous.

Comments: Galápagos shore petunia is seen along coastlines throughout the archipelago. Occasionally it is found inhabiting the arid lowlands as well. This species is referred to as *Cacabus miersii* (Hook. f.) Wettst. in Wiggins and Porter (1971).

235. *Exedeconus miersii* (Galápagos shore petunia), flowers

Scientific Name: Solanum americanum Mill. (Photo 236)
Common Names: Glossy nightshade, *hierba mora*

Family: Solanaceae (Potato)

Range: Native; also known from other tropical regions throughout the world.

Islands Inhabited: Fernandina, Floreana, Isabela (A,CA,D,SN), Pinta, Pinzón, San Cristóbal, Santa Cruz, Santiago, Wolf

Habitat: Arid lowlands

Description: Erect or spreading perennial herb to 60 cm tall, occasionally somewhat woody at the base. *Leaves* alternate, simple, ovate to ovate-lanceolate, 2–14 cm long, margins entire, wavy, or toothed. *Flowers* in lateral, umbel-like cymes; corolla white, sometimes tinged with purple, rotate with 5 lobes, 4–5 mm across; stamens 5, anthers yellow. *Fruit* a berry, dark purple to black, round, ca. 5 mm in diameter, fleshy; seeds numerous.

Comments: Glossy nightshade, referred to as *Solanum nodiflorum* Jacq. in Wiggins and Porter (1971), is common to disturbed areas such as fields, roadsides, and around buildings.

Solanum melongena L., commonly called "eggplant," is cultivated on San Cristóbal and Santa Cruz. *Solanum tuberosum* L., the Irish or white potato, is cultivated on Santa Cruz.

236. *Solanum americanum* (glossy nightshade), flowers and fruits

See also

Adenostemma platyphyllum (p. 247)
Blainvillea dichotoma (p. 248)
Desmodium limense (p. 223)
D. procumbens (p. 223)
Heliotropium indicum (p. 219)
Polygala anderssonii (p. 227)
P. galapageia (p. 227)
P. sancti-georgii (p. 227)
Pseudelephantopus spicatus (p. 218)
P. spiralis (p. 218)
Sida acuta (p. 206)
S. rhombifolia (p. 206)
S. salviifolia (p. 206)

Herbs with Alternate Leaves and Yellow or Orange Flowers

Scientific Name: Sonchus oleraceus L. (Photo 237)
Common Name: Sow thistle

Family: Asteraceae (Sunflower)

Range: Introduced; also known from other tropical regions throughout the world, originally from Europe.

Islands Inhabited: Fernandina, Floreana, Isabela (A,CA,D,E,SN,W), San Cristóbal, Santa Cruz, Santiago

Habitat: Arid lowlands and moist uplands

Description: Annual herb to ca. 1.5 m tall, stems containing a milky sap. *Leaves* alternate, simple, deeply divided or sometimes simply toothed, lanceolate to spatulate, 6–30 cm long, base clasping the stem, margins prickly. *Flowers* in ligulate heads 1.5–2.5 cm across; heads numerous; flowers yellow, usually 120–60 per head, perfect. *Fruit* an achene, flattened, to 4 mm long, numerous ribs visible on the surface, numerous hairlike bristles on top, to 8 mm long; seed 1.

Comments: Sow thistle is a common weed of disturbed areas such as roadsides and gardens. Its bright yellow flowers are often visited by the Galápagos carpenter bee (*Xylocopa darwini*).

237. *Sonchus oleraceus* (sow thistle), flowers and fruits

Scientific Name: Senna obtusifolia (L.) Irwin & Barneby (Photo 238)
Common Name: Sicklepod

Family: Caesalpiniaceae (Caesalpinia)

Range: Introduced; also known from other tropical regions throughout the world.

Islands Inhabited: Floreana, Isabela (W), San Cristóbal, Santa Cruz, Santiago

Habitat: Arid lowlands

Description: Herb to 1 m tall, woody at the base. *Leaves* alternate, even-pinnately compound; leaflets 4–6, oval to obovate, tips somewhat rounded, middle ones 2.5–5.5 cm long, covered with tiny white hairs, margins fringed with tiny hairs. *Flowers* solitary or in axillary racemes; corolla yellow, petals 5, 9–12 mm long; fertile stamens 7 (3 large, 4 small). *Fruit* a legume, curved and 4-angled, 10–20 cm long; seeds numerous.

Comments: This species is referred to as *Cassia tora* L. in Wiggins and Porter (1971). In certain parts of the world, the fruits and seeds have been used medicinally, and the seeds have been roasted as a coffee substitute.

A similar species, *Senna uniflora* (P. Miller) Irwin & Barneby (rufous-haired senna), listed as *C. uniflora* Mill. in Wiggins and Porter (1971), differs in having reddish brown hairs covering most of its parts. It is a native and occurs on Baltra, San Cristóbal, and Santa Cruz.

238. *Senna obtusifolia* (sicklepod), flowers and fruits

Scientific Name: Senna occidentalis (L.) Link (Photo 239)
Common Name: Coffee senna

Family: Caesalpiniaceae (Caesalpinia)

Range: Native; also known from other tropical regions throughout the world.

Islands Inhabited: Floreana, Isabela (SN), San Cristóbal, Santa Cruz, Santiago

Habitat: Arid lowlands

Description: Perennial herb to 2 m tall, woody at the base. *Leaves* alternate, even-pinnately compound; leaflets 8–12 (usually 10), elliptic, ovate or lanceolate, sharp-pointed, to ca. 6 cm long, margins fringed with tiny hairs. *Flowers* in axillary or terminal racemes; corolla yellow, petals 5, 12–15 mm long; fertile stamens 6 (2 large, 4 small). *Fruit* a legume, linear, 8–10 cm long; seeds numerous, grayish green.

Comments: Coffee senna, listed as *Cassia occidentalis* L. in Wiggins and Porter (1971), appears to favor open, disturbed areas. It is often seen along roadsides and in fields. In certain parts of the world, it has been used as a coffee substitute, although it does not contain caffeine. It is also used medicinally.

Senna hirsuta (L.) Irwin & Barneby var. *hirsuta* (hairy senna), listed as *C. hirsuta* L. in Wiggins and Porter (1971), also has leaflets that are sharp-pointed. However, this introduced weed, known only from Floreana, is densely covered with hairs, while *S. occidentalis* is relatively smooth. Hairy senna may also be recognized by its foul odor.

239. *Senna occidentalis* (coffee senna), flower

Scientific Name: Cleome viscosa L. (Photo 240)
Common Names: Cleome

Family: Capparidaceae (Caper)

Range: Introduced; also known from other tropical regions throughout the world, originally probably from tropical Asia.

Islands Inhabited: Baltra, Floreana, Pinta, Santa Cruz, Santiago, Seymour

Habitat: Arid lowlands

Description: Erect annual herb to 1.6 m tall, covered with sticky hairs. *Leaves* alternate, palmately compound; leaflets 3–5, oblanceolate-elliptic, to 6 cm long, covered with sticky hairs, margins entire. *Flowers* axillary, solitary; corolla yellow, petals 4, 6–14 mm long; stamens 10–26. *Fruit* a capsule, linear-cylindrical, 3–10 cm long, covered with sticky hairs; seeds numerous, dark reddish brown.

Comments: This introduced weed was first noticed in 1963 on the island of Baltra (Porter 1984). Since that time it has spread to five additional islands, most recently Seymour. This, no doubt, is due to the ease with which its sticky fruits are dispersed.

240. *Cleome viscosa* (cleome), flowers and fruits

Scientific Name: Crotalaria incana L. (Photo 241)
Common Name: Fuzzy rattlebox

Family: Fabaceae (Pea)

Range: Native; also known from other tropical regions throughout the world, originally from tropical America.

Islands Inhabited: Fernandina, Isabela (A,D,SN), Marchena, Pinta, San Cristóbal, Santa Cruz, Santiago

Habitat: Arid lowlands and moist uplands

Description: Erect perennial herb to ca. 2 m tall, older stems becoming somewhat woody. *Leaves* alternate, odd-pinnately compound; leaflets 3, elliptic, ovate, roundish, or obovate, 1–5 cm long. *Flowers* in racemes, opposite the leaves or terminal; corolla yellow, flower ca. 1.5–1.8 cm long, composed of 1 large standard petal (usually tinged reddish to reddish brown), 2 lateral wing petals, and 2 lower keel petals that are somewhat fused; stamens 10, fused into a tube. *Fruit* a legume, tan or brownish, linear-oblong, 2–5 cm long, inflated; seeds numerous, yellow to brownish.

Comments: Wiggins and Porter (1971) list two varieties for *Crotalaria incana*. The young stems, leaflets, and fruits of var. *incana* (Photo 241) are typically hairy, while those of var. *nicaraguensis* Senn. are smooth. The former occurs on Fernandina, Isabela (D,SN), Marchena, Pinta, Santa Cruz, and Santiago. The latter is known from Fernandina, Isabela (D), San Cristóbal, and Santa Cruz. Some taxonomists do not believe that the above differences merit recognition at the varietal rank. This species is normally found in disturbed sites such as roadsides and fields.

Crotalaria pumila Gomez Ortega (dwarf rattlebox) is similar to *C. incana* but has smaller leaflets (0.5–3.5 cm long), flowers (ca. 0.6–1.2 cm long), and fruits (1–2 cm long). This native is found on Baltra, Fernandina, Floreana, Isabela (A,D,SN), Pinta, Pinzón, Rábida, San Cristóbal, Santa Cruz, Santa Fe, Santiago, Seymour, and one or more of the smaller islands.

241. *Crotalaria incana* var. *incana* (fuzzy rattlebox), flowers and fruits

Scientific Name: Crotalaria retusa L. (Photos 242–43)
Common Name: Rattlebox

Family: Fabaceae (Pea)

Range: Introduced; also known from other tropical regions throughout the world, originally from Asia.

Islands Inhabited: San Cristóbal, Santa Cruz

Habitat: Arid lowlands

Description: Erect annual herb to ca. 1 m tall, young stems somewhat hairy. *Leaves* alternate, simple, oblanceolate, to ca. 8 cm long, margins entire. *Flowers* in terminal racemes; corolla yellow, flower to ca. 2.5 cm long, composed of 1 large standard petal (usually tinged reddish to reddish brown), 2 lateral wing petals, and 2 lower keel petals that are somewhat fused; stamens 10, fused into a tube. *Fruit* a legume, dark brown to black, linear-oblong, 3–4 cm long, inflated; seeds numerous, tan to brownish.

Comments: In some parts of the world, this species is cultivated as an ornamental, but in the Galápagos it is normally found growing in disturbed sites. The genus name is derived from the Greek *krotalon,* which means "castanet." This refers to the sound made by the seeds when the mature, dried fruit is shaken.

242. (above) *Crotalaria retusa* (rattlebox), flowers and fruits

243. (left) *Crotalaria retusa* (rattlebox), fruits

Scientific Name: Vigna luteola (Jacq.) Benth. (Photo 244)
Common Name: Wild cowpea

Family: Fabaceae (Pea)

Range: Native; also known from other tropical regions throughout the world.

Islands Inhabited: Fernandina, Isabela (A,SN), Pinta, Santa Cruz

Habitat: Moist uplands

Description: Spreading or climbing perennial herb. *Leaves* alternate, odd-pinnately compound; leaflets 3, ovate to lanceolate, ca. 2–9 cm long. *Flowers* in axillary racemes; corolla yellow, flower 1.5–2 cm long, composed of 1 large standard petal, 2 lateral wing petals, and 2 lower keel petals that are somewhat fused; stamens 10, 9 of these fused into a tube, the other free. *Fruit* a legume, oblong, 4–7 cm long; seeds few to numerous, black.

Comments: The genus name honors the seventeenth-century Italian botanist Dominico Vigna. This plant is similar in many ways to the various species of *Phaseolus*; however, no Galápagos members of the latter genus have yellow flowers.

244. *Vigna luteola* (wild cowpea), flower

Scientific Name: Mentzelia aspera L. (Photo 245)
Common Name: Stickleaf

Family: Loasaceae (Stickleaf)

Range: Native; also known from other regions of tropical America (Mexico and the West Indies to the northern half of South America).

Islands Inhabited: Baltra, Daphne, Española, Floreana, Genovesa, Isabela (CA,D,SN), Pinta, Pinzón, Rábida, San Cristóbal, Santa Cruz, Santa Fe, Santiago, Islet(s)

Habitat: Arid lowlands

Description: Spreading annual herb to ca. 40 cm tall, stems covered with barbed hairs. *Leaves* alternate, simple; blade sometimes 3-lobed, ovate, ca. 1–6 cm long, both surfaces covered with barbed hairs, margins dentate. *Flowers* axillary, solitary; corolla yellow, petals 5, 3–6 mm long; stamens ca. 20. *Fruit* a capsule, somewhat cone-shaped, 5–11 mm long, covered with barbed hairs; seeds 1 to numerous.

Comments: The barbed hairs on the leaves and fruits of this plant make them tremendously successful at attaching to any animals that pass by, including humans. After a day's walk, one is sure to have several of these "hitchhikers" firmly attached to pants, socks, or shoes. Thus, stickleaf is extremely widespread in the archipelago. The genus name honors the German botanist Christian Mentzel (1622–1701).

245. *Mentzelia aspera* (stickleaf), flower and fruits

Scientific Name: Bastardia viscosa (L.) HBK. (Photo 246)
Common Name: Bastardia

Family: Malvaceae (Mallow)

Range: Native; also known from other regions of tropical America.

Islands Inhabited: Española, Floreana, Isabela (A,CA,SN), Pinta, Pinzón, Rábida, San Cristóbal, Santa Cruz, Santiago, Seymour

Habitat: Arid lowlands

Description: Perennial herb to 1 m tall, somewhat woody at the base, covered with various hairs, some glandular. *Leaves* alternate, simple; blade ovate to somewhat cordate, to 10 cm long, margins entire or with small teeth. *Flowers* axillary, solitary or in threes; corolla yellow to yellowish orange, petals 5, 5–6 mm long; stamens numerous, united in a tubular column. *Fruit* a schizocarp, roundish, ca. 6–7 mm across, usually with 6–7 sections, beaked, seed 1 per section, brown, with silky hairs.

Comments: This herb is often found growing in shrubby areas that receive abundant sunlight.

246. *Bastardia viscosa* (bastardia), flowers and fruits

Scientific Name: Sida rhombifolia L. (Photo 247)
Common Names: Sida, escoba

Family: Malvaceae (Mallow)

Range: Introduced; also known from other tropical regions throughout the world.

Islands Inhabited: Floreana, Isabela (A,SN), San Cristóbal, Santa Cruz

Habitat: Moist uplands

Description: Erect perennial herb to ca. 1.5 m tall, much-branched, older stems becoming woody. *Leaves* alternate, simple; blade rhombic to elliptic or oblong-ovate, typically 2–6 cm long, sometimes longer, margins serrate toward the tip, entire toward the base. *Flowers* axillary, solitary or few-flowered, occasionally appearing as a terminal cluster; corolla yellow or orange-yellow to white, sometimes with brownish red or purple markings at the base, petals 5, 7–10 mm long, sometimes longer; stamens numerous, united in a tubular column, yellow; style branches brownish red. *Fruit* a schizocarp, depressed ovate, 5–6 mm across, with 7–14 sections, each with 2 short awns on top; seed 1 per section, brown.

Comments: This species appears to favor open, disturbed sites. It is one of the favorite food plants for Galápagos tortoises in the wild. Most other members of the genus are relatively difficult to distinguish. However, *Sida glutinosa* Cav., an introduced weed on Isabela (A) and Santa Cruz, is different from all other species in having glandular hairs (van der Werff 1977). Its corolla is pale orange in color. *Sida rupo* Ulbr., a native inhabiting Isabela (A,CA), is unique in having palmately lobed leaves. Its petals are purplish.

Fruits, leaves, and flower color are the primary characters used to identify the remaining species. For example, *S. acuta* Burm. f. (introduced) and *S. salviifolia* C. Presl (native) produce fruits with 6 to 7 sections. They differ in that *S. acuta* has oblong-ovate to rhombic-shaped leaves, while *S. salviifolia* has leaves that are usually linear to linear-lanceolate. *Sida acuta* (corolla white to yellow) is found on Floreana and Isabela (A,CA,SN), while *S. salviifolia* (corolla whitish) is known from Daphne, Española, Fernandina, Floreana, Genovesa, Isabela (A,D,SN), Pinta, Rábida, San Cristóbal, Santa Cruz, Santa Fe, Santiago, Seymour, and one or more of the smaller islands.

Sida spinosa L. (native), *S. hederifolia* Cav. (native), and *S. paniculata* L. (introduced) produce fruits with only 5 sections. *Sida spinosa* has linear to narrowly ovate leaves, while *S. hederifolia* and *S. paniculata* have leaves that are ovate to almost roundish. The latter two species differ in that *S. hederifolia* has a yellowish orange corolla that is reddish at the base, while *S. paniculata* has a brownish red to purple corolla. *Sida spinosa* (corolla yellow to yellowish orange) occupies Española, Floreana, Isabela (A,CA,SN), Pinta, Pinzón, San Cristóbal, Santa Cruz, and Santiago. *Sida hederifolia* occurs on Floreana, Isabela (A,CA,D,SN), Pinta, Pinzón, San Cristóbal, Santa Cruz, and Santiago. *Sida paniculata* is found on Floreana, Isabela (CA), and San Cristóbal.

Sida veronicifolia Lam., listed as *S. veronicaefolia* Lam. in Hamann (1979), is a native known only from Pinta. It is similar to *S. hederifolia*. However, *S. veronicifolia* has leaf blades that are ca. 3–5 cm long, while *S. hederifolia* has leaf blades that are typically less than 2.5 cm long (Hamann 1979).

247. *Sida rhombifolia* (sida), flower

Scientific Name: Ludwigia leptocarpa (Nutt.) Hara (Photo 248)
Common Name: False loosestrife

Family: Onagraceae (Evening-primrose)

Range: Native; also known from other tropical regions throughout the world.

Islands Inhabited: San Cristóbal, Santa Cruz

Habitat: Moist uplands

Description: Erect perennial herb to ca. 1 m tall, much-branched, stems reddish. *Leaves* alternate, simple, broadly lanceolate, 3.5–18 cm long, splotched with red, margins usually entire. *Flowers* axillary and solitary or in terminal clusters; corolla yellow, petals usually 5, 5–11 mm long; stamens 10. *Fruit* a capsule, brown at maturity, cylindrical with 10 ribs, 1.5–5 cm long; seeds numerous.

Comments: This species typically inhabits moist areas. The genus name honors Christian Ludwig (1709–73) of Leipzig, Germany. *Ludwigia peploides* (HBK.) Raven subsp. *peploides* also has 5 petals but differs from *L. leptocarpa* in that it is typically a floating plant. It is native and occurs on Isabela (SN), San Cristóbal, Santa Cruz, and Santiago.

Ludwigia erecta (L.) Hara, also native, differs from both of the above-mentioned species in having 4 petals rather than 5. It occurs on Española, Floreana, San Cristóbal, and Santa Cruz.

248. *Ludwigia leptocarpa* (false loosestrife), flower and fruits

Scientific Name: Oxalis corniculata L. (Photo 249)
Common Name: Creeping wood sorrel

Family: Oxalidaceae (Oxalis)

Range: Introduced; also known from other tropical regions throughout the world.

Islands Inhabited: Floreana, Isabela (A,SN), San Cristóbal, Santa Cruz

Habitat: Moist uplands

Description: Low-growing annual herb, creeping, stems rooting at the nodes. *Leaves* alternate, palmately compound; leaflets 3, obcordate, 5–11 mm long, 8.5–18 mm wide, margins fringed with fine hairs. *Flowers* in axillary, umbel-like cymes; corolla yellow, sometimes orange along the veins, petals 5, 6 mm long; stamens 10. *Fruit* a capsule, cylindrical, 1–1.6 cm long; seeds numerous, brown.

Comments: Oxalis megalorrhiza Jacq. and *O. dombeyi* (Dombey's wood sorrel) also have yellow flowers. The former, a native found on Floreana, Pinta, Pinzón, and Santa Cruz, differs from *O. corniculata* in not having creeping stems that root at the nodes. The latter, also a native, differs from *O. corniculata* and *O. megalorrhiza* in having pinnately rather than palmately compound leaves.

249. *Oxalis corniculata* (yellow wood sorrel), flowers

Scientific Name: Oxalis dombeyi A. St.-Hil. (Photo 250)
Common Name: Dombey's wood sorrel

Family: Oxalidaceae (Oxalis)

Range: Native; also known from Panama, mainland Ecuador, and Peru.

Islands Inhabited: Española, Fernandina, Floreana, Isabela (CA,D), Pinzón, Rábida, San Cristóbal, Santa Cruz, Santa Fe, Santiago, Seymour, Islet(s)

Habitat: Arid lowlands and moist uplands

Description: Annual herb to 50 cm tall, becoming slightly woody at the base, stems sometimes reddish or brownish, with glandular hairs. *Leaves* alternate, pinnately compound; leaflets 3, terminal leaflet typically obovate, 5–23 mm long, 6–24 mm wide, stalk holding the terminal leaflet is 2–6 mm long, lateral leaflets somewhat obovate, 5–17 mm long, 3.5–11 mm wide, margins fringed with fine hairs. *Flowers* in axillary, umbel-like cymes; corolla yellow, sometimes orange along the veins, petals 5, 5.5–7 mm long; stamens 10. *Fruit* a capsule, oblong, 7–10 mm long; seeds numerous, reddish.

Comments: Dombey's wood sorrel is referred to as *O. cornellii* Andersson in Wiggins and Porter (1971). The genus name comes from the Greek *oxys*, meaning "acid." This refers to the oxalic acid produced by the plant, which causes a sour taste when the leaves are chewed.

250. *Oxalis dombeyi* (Dombey's wood sorrel), flower and fruit

Scientific Name: Portulaca howellii (D. Legrand) Eliasson (Photo 251)
Common Names: Galápagos purslane

Family: Portulacaceae (Purslane)

Range: Endemic

Islands Inhabited: Daphne, Darwin, Floreana, Genovesa, Marchena, Pinta, Plaza Sur, Rábida, Santa Fe, Seymour, Sombrero Chino, Wolf, Islet(s)

Habitat: Arid lowlands

Description: Low-growing herb, occasionally forming mats, stems fleshy, to 1 cm thick, sometimes reddish. *Leaves* alternate or almost opposite, simple, wedge-shaped to obovate, to 3 cm long, fleshy, margins entire. *Flowers* solitary; tepals 5, yellow, typically to 1.5 cm long, ocasionally longer; stamens 5 to numerous. *Fruit* a capsule, 7–8 mm long, opening by means of a lid; seeds numerous, black, covered with rows of tiny bumps.

Comments: Two additional species of *Portulaca* have yellow flowers. These are *P. oleracea* (common purslane) and *P. umbraticola*. They differ from *P. howellii* in that their flowers are distinctly smaller and their stems are thinner. In addition, their leaves are persistent, while those of *P. howellii* are short-lived. Thus, *P. howellii* is leafless most of the year.

251. *Portulaca howellii* (Galápagos purslane), flowers

Scientific Name: Portulaca oleracea L. (Photo 252)
Common Names: Common purslane, verdolaga

Family: Portulacaceae (Purslane)

Range: Native; also known from other tropical regions throughout the world, probably originally from India.

Islands Inhabited: Española, Fernandina, Floreana, Isabela (A,CA,SN), Pinta, San Cristóbal, Santa Cruz, Santiago, Wolf, Islet(s)

Habitat: Arid lowlands

Description: Low-growing annual herb, much-branched and forming mats, stems fleshy, sometimes reddish. *Leaves* alternate or almost opposite, simple, wedge-shaped to obovate, 1–2.5 cm long, fleshy, margins entire. *Flowers* in axillary clusters; tepals 5, yellow, 3–8 mm long; stamens 5 to numerous. *Fruit* a capsule, ca. 4 mm long, opening by means of a thin, translucent lid; seeds numerous, black, shiny, covered with rows of tiny bumps.

Comments: This species is common in disturbed areas such as roadsides, gardens, and yards. *Portulaca umbraticola* HBK., also a native, differs primarily in having a capsule with a thick, nontranslucent lid and gray seeds. It occurs on Española, Fernandina, Isabela (CA,D,W), Pinzón, Rábida, Santa Cruz, Santiago, and one or more of the smaller islands.

Portulaca grandiflora Hook. (rose-moss), an ornamental, may be seen in the town of Puerto Ayora (Santa Cruz). It differs from the other members of the genus in having large, reddish to purplish flowers.

252. *Portulaca oleracea* (common purslane), flower

Scientific Name: Lycopersicon cheesmanii Riley (Photos 253–54)
Common Names: Galápagos tomato, *tomatillo*

Family: Solanaceae (Potato)

Range: Endemic

Islands Inhabited: Baltra, Darwin, Española, Fernandina, Floreana, Isabela (A,CA,D, E,SN,W), Pinta, Pinzón, Rábida, San Cristóbal, Santa Cruz, Santa Fe, Santiago, Islet(s)

Habitat: Arid lowlands

Description: Perennial herb, much-branched, stems sprawling over the ground, covered with short hairs. *Leaves* alternate, odd-pinnately compound, 4–12 cm long; leaflets variable in number, margins crenate. *Flowers* in terminal and lateral racemes, sometimes forked; corolla yellow, rotate with 5 lobes, 1–1.2 cm across, lobes bending backward; stamens 5, anthers yellow, united into a column, 4–5 mm long. *Fruit* a berry, golden yellow or orange to deep red, round, 8–12 mm in diameter, fleshy; seeds numerous.

Comments: This species includes two varieties. *Lycopersicon cheesmanii* var. *cheesmanii* (Photo 253) is found on Baltra, Darwin, Española, Fernandina, Floreana, Isabela (A,CA,E,SN,W), Pinta, Pinzón, San Cristóbal, Santa Cruz, Santa Fe, and Santiago. It is frequently found on rough lava areas as well as disturbed sites, such as roadsides. Although not cultivated as a food crop, the fruits have a delicious, tangy flavor.

Lycopersicon cheesmanii var. *minor* (Hook. f.) D. M. Porter (hairy Galápagos tomato) (Photo 254), referred to as *L. cheesmanii* Riley forma *minor* Muller in Wiggins and Porter (1971), occurs on Fernandina, Isabela (A,CA,D,SN), Pinta, Pinzón, Rábida, Santiago, and one or more of the smaller islands. It differs from var. *cheesmanii* in having stems and leaves that are much hairier, a distinctive acrid odor, leaves that are more highly divided, and a fruit wall that is thinner. Studies by Rick and Bowman (1961) demonstrated that passage through the gut of a tortoise increased the frequency of seed germination in var. *minor*. They suggested that tortoises use tomatoes as a food source and in turn act as a dispersal agent. As seeds may remain in a tortoise's digestive tract for up to four weeks, this assures dissemination far from the parent plant.

Lycopersicon esculentum Mill. is the common garden tomato. It is cultivated on San Cristóbal and Santa Cruz.

253. *Lycopersicon cheesmanii* var. *cheesmanii* (Galápagos tomato), flowers and fruits

254. *Lycopersicon cheesmanii* var. *minor* (hairy Galápagos tomato), flowers

Scientific Name: Physalis pubescens L. (Photo 255)
Common Name: Hairy ground cherry

Family: Solanaceae (Potato)

Range: Native; also known from other tropical regions throughout the world, originally from tropical America.

Islands Inhabited: Baltra, Española, Fernandina, Genovesa, Isabela (A,CA,D,SN), Pinta, Pinzón, Rábida, Santa Cruz, Santiago, Seymour

Habitat: Arid lowlands

Description: Annual herb to 90 cm tall, stems hairy and somewhat sticky. *Leaves* alternate, simple; blade ovate, 4–9 cm long, both surfaces somewhat hairy, margins irregulary toothed, occasionally entire. *Flowers* axillary, solitary; corolla yellow with dark splotches in the throat, campanulate with 5 lobes, typically 7–10 mm long, throat hairy below the dark splotches; stamens 5, anthers bluish. *Fruit* a berry, round, 10–18 mm in diameter, enclosed in an inflated, 5-angled calyx, 18–30 mm long and 13–22 mm across, covered with hairs; seeds numerous.

Comments: This plant grows in a variety of locations, from open, disturbed sites to forested areas. The genus name comes from the Greek *physa*, "bladder," which refers to the inflated, bladderlike calyx.

Physalis galapagoensis Waterfall (Galápagos ground cherry), an endemic, lacks the above-mentioned dark splotches in its corolla throat. In addition, the inflated calyx is 30–45 mm long and 37–40 mm across and is smooth rather than hairy. This plant inhabits Española, Fernandina, Floreana, Isabela (D,SN), Rábida, San Cristóbal, Santa Cruz, Seymour, and one or more of the smaller islands.

Physalis angulata L. (ground cherry), a native, differs from both of these in having a 10-angled, inflated calyx. It occurs on Española, Floreana, Isabela (D,SN), Marchena, San Cristóbal, Santa Cruz, and Santiago.

Physalis peruviana L. (cape gooseberry), cultivated for its edible fruits, is much larger than the other varieties (van der Werff 1977). It is known only from Isabela (SN) and Santa Cruz.

255. *Physalis pubescens* (hairy ground cherry), flowers and fruit

See also

Capsicum frutescens (p. 194)
Heliotropium anderssonii (p. 180)

Herbs with Alternate Leaves and Pink, Red, or Purple Flowers

Scientific Name: Elytraria imbricata (M. Vahl) Pers. (Photos 256–57)
Common Name: Elytraria

Family: Acanthaceae (Acanthus)

Range: Introduced; also known from other tropical regions throughout the world.

Islands Inhabited: Santa Cruz

Habitat: Arid lowlands

Description: Herb to 60 cm tall. *Leaves* alternate, clustered near the stem tips, simple; blade oblong, ovate or obovate, to 18 cm long, margins entire. *Flowers* in spikes to 6 cm long, covered with tiny bracts; corolla purplish to blue, tubular with 2 distinct lips, 5–8 mm long, upper lip 2-lobed, lower lip 3-lobed; stamens 2. *Fruit* a capsule, oblong, to 5 mm long; seeds numerous.

Comments: This plant, with its small flowers, is often overlooked. However, a nice population exists at the lava tunnel near Puerto Ayora.

256. (above) *Elytraria imbricata* (elytraria)

257. (left) *Elytraria imbricata* (elytraria), flower

Scientific Name: Porophyllum ruderale (Jacq.) Cass.

(Photos 258–59, see also Photo 12)

Common Name: Poreleaf

Family: Asteraceae (Sunflower)

Range: Introduced; also known from other regions of tropical America (southwestern United States to northern South America).

Islands Inhabited: Daphne, Española, Floreana, Isabela (A,D,SN,W), Pinta, Pinzón, Rábida, San Cristóbal, Santa Cruz, Santiago, Seymour, Islet(s)

Habitat: Arid lowlands

Description: Annual herb to 1 m tall, stems with a bluish, waxy covering. *Leaves* alternate or opposite, simple; blade broadly elliptic, 1.5–4 cm long, with a bluish, waxy covering, oil glands present, veins sometimes purple, margins wavy, occasionally toothed, sometimes purple. *Flowers* in discoid heads, disc 6–7 mm across; heads solitary, usually terminal; flowers purplish or greenish, ca. 21–65 or more per head, perfect, style branches bright yellow. *Fruit* an achene, 1–1.5 cm long, with numerous hairlike bristles on top, to 1 cm long; seed 1

Comments: All Galápagos members of this species represent var. *macrocephalum* (DC.) Cronquist. As the common name suggests, this plant's leaves are covered with small glands or "pores." These leaves produce an unpleasant odor when crushed.

258. (above) *Porophyllum ruderale* var. *macrocephalum* (poreleaf)

259. (left) *Porophyllum ruderale* var. *macrocephalum* (poreleaf), flowers and fruits

Scientific Name: Pseudelephantopus spiralis (Less.) Cronquist (Photo 260)
Common Name: False elephant's foot

Family: Asteraceae (Sunflower)

Range: Introduced; also known from other regions of tropical America.

Islands Inhabited: Isabela (SN), San Cristóbal, Santa Cruz

Habitat: Moist uplands

Description: Perennial herb to 1 m tall, stems extremely hairy, especially younger growth, spreading by runners. *Leaves* alternate, simple, lower leaves 5–25 cm long (including petiole), upper leaves smaller; petiole often somewhat winged; blade elliptic to oblanceolate or obovate, both surfaces somewhat hairy, margins serrate. *Flowers* in discoid heads; heads solitary or in small clusters on spikes 5–25 cm long; flowers purplish or blue to white, 4 per head, perfect. *Fruit* an achene, 3–4 mm long, several principal awns on top with upper parts curled or kinked, 3–5 mm long; seed 1.

Comments: Pseudelephantopus spicatus (Juss.) C. F. Baker differs from *P. spiralis* in having longer achenes (5.5–8 mm) and longer principal awns (5–7 mm), of which there are 2. One or both of these awns may be sharply bent in a double-reversed fashion. This introduced species is also found on Isabela (SN), San Cristóbal, and Santa Cruz.

260. *Pseudephantopus spiralis* (false elephant's foot), flowers

Scientific Name: Heliotropium indicum L. (Photo 261)
Common Names: Indian heliotrope, *cola de escorpion*

Family: Boraginaceae (Borage)

Range: Native; also known from other tropical regions throughout the world.

Islands Inhabited: Floreana, San Cristóbal, Santa Cruz

Habitat: Moist uplands

Description: Herb to 1 m tall, much-branched, covered with hairs. *Leaves* alternate, simple; blade ovate to broadly lanceolate, 3–15 cm long, hairy, margins entire to slightly wavy. *Flowers* in scorpioid spikes 10–30 cm long; corolla typically purplish or bluish with a white and/or yellow throat, occasionally entirely white, funnelform with 5 lobes, 3–5 mm long; stamens 5. *Fruit* consisting of 4 nutlets that separate at maturity, 2.5–3.5 mm long, each with a pointed tip and 2–3 lines on the back; seed 1 per nutlet.

Comments: Indian heliotrope is easily distinguished from other members of the genus by its pale purplish to bluish flowers. It appears to prefer open areas with a mixture of sunshine and shade. The genus name is derived from the Greek *helios*, "sun," and *trope*, "to turn." This is based on an ancient but incorrect belief that the flowers follow the sun as it moves across the sky.

261. *Heliotropium indicum* (Indian heliotrope), flowers

Scientific Name: Tradescantia zebrina Bosse (Photo 262)
Common Name: Wandering Jew

Family: Commelinaceae (Spiderwort)

Range: Cultivated escape; also known from other regions of tropical America, originally from Mexico and Guatemala.

Islands Inhabited: San Cristóbal, Santa Cruz

Habitat: Moist uplands

Description: Low-growing perennial herb, stems purplish green. *Leaves* alternate, simple; petiole forming a sheath on the stem; blade ovate-oblong, ca. 2–8 cm long, somewhat fleshy, upper surface striped purple and green, lower surface purple, margins entire except for sheath, fringed with fine white hairs. *Flowers* in terminal or axillary cymes, each cyme or pair of cymes with 2 bracts beneath; corolla reddish purple, tubular with 3 lobes, lobes ca. 5 mm long; stamens 6. *Fruit* a capsule; seeds 1–6.

Comments: The stems of wandering Jew are somewhat slimy and sticky when crushed. In addition, its petals are so delicate that they seem to disappear when rubbed between the fingers. This plant is often cultivated as a garden and yard ornamental or as a house plant and is easily propagated by cuttings.

262. *Tradescantia zebrina* (wandering Jew), flower

Scientific Name: Evolvulus convolvuloides (Willd. ex J. A. Schult.) Stearn (Photo 263)
Common Name: Purple evolvulus

Family: Convolvulaceae (Morning-glory)

Range: Native; also known from other regions of tropical America (Mexico and Florida to central South America).

Islands Inhabited: Española, Floreana, Isabela (A,CA,SN), Pinta, Pinzón, Rábida, San Cristóbal, Santa Cruz, Santa Fe, Santiago, Seymour, Islet(s)

Habitat: Arid lowlands

Description: Prostrate herb to ca. 50 cm long, sometimes longer, much-branched. *Leaves* alternate, simple; blade broadly elliptic to somewhat roundish, 6–20 mm long, margins entire. *Flowers* axillary, solitary; corolla purplish or bluish, somewhat funnelform-rotate, ca. 4 mm long, 7–10 mm across; stamens 5. *Fruit* a capsule, broadly obovoid to somewhat spherical, ca. 3 mm long; seeds 1–4, dark brown.

Comments: This species is listed as *E. glaber* Spreng. in Wiggins and Porter (1971). The genus name comes from the Latin *evolvere*, "to unwind." This alludes to the plant's nontwining form as compared with most other members of the family Convolvulaceae.

263. *Evolvulus convolvuloides* (blue evolvulus), flowers

Scientific Name: Clitoria ternatea L. (Photo 264)
Common Name: Butterfly pea

Family: Fabaceae (Pea)

Range: Cultivated escape; also known from other tropical regions throughout the world.

Islands Inhabited: San Cristóbal, Santa Cruz

Habitat: Arid lowlands

Description: Twining perennial herb, older stems becoming slightly woody. *Leaves* alternate, odd-pinnately compound; leaflets 5–7 (occasionally 9), elliptic or oblong, to ca. 5 cm long. *Flowers* axillary, solitary or paired; corolla purple to dark blue with yellowish white in the center, composed of 1 large standard petal (3–5 cm long), 2 lateral wing petals, and 2 lower keel petals that are somewhat fused; stamens 10, 9 of these fused into a tube, the other free. *Fruit* a legume, linear-oblong, to 12 cm long; seeds 8–10, dark brown.

Comments: The large, showy flowers of this plant are distinctive and should not be confused with any other Galápagos members of the pea family.

264. *Clitoria ternatea* (butterfly pea), flowers and fruits

Scientific Name: Desmodium incanum DC. (Photo 265)
Common Names: Tick trefoil

Family: Fabaceae (Pea)

Range: Native; also known from other tropical regions throughout the world, originally from tropical America.

Islands Inhabited: Fernandina, Floreana, Genovesa, Isabela (D,SN), San Cristóbal, Santa Cruz

Habitat: Arid lowlands and moist uplands

Description: Erect perennial herb to ca. 1 m tall, occasionally somewhat woody at the base. *Leaves* alternate, odd-pinnately compound; leaflets 3, elliptic to ovate, 2–11 cm long, lower surface with hairs. *Flowers* in terminal or axillary racemes; corolla pinkish to purplish, flower ca. 5 mm long, composed of 1 long standard petal, 2 lateral wing petals, and 2 lower keel petals that are somewhat fused; stamens 10, 9 of these fused into a tube, the other free. *Fruit* a loment, not twisted, composed of ca. 3–9 segments, each somewhat roundish, 3.5–4 mm long, 2.5 mm wide; seeds 1 per segment.

Comments: The genus name of this species, listed as *Desmodium canum* (J. F. Gmel.) Schinz & Thell. in Wiggins and Porter (1971), comes from the Greek *desmos*, which means "chain." This refers to the individual segments making up each fruit.

Desmodium limense Hook. is similar to *D. incanum* in that its fruits are not twisted, but its flowers are usually larger (6–10 mm long), and its fruit segments are somewhat rhomboid-shaped. Its corolla may be pinkish, purplish, bluish, or white. This native occurs on Fernandina, Isabela (A,D), San Cristóbal, Santa Cruz, and Santiago.

Desmodium glabrum (Mill.) DC. and *D. procumbens* (Mill.) Hitchc., both natives, differ from the above-mentioned species in that their flowers are smaller and their fruits are slightly twisted. *Desmodium glabrum* (corolla pinkish or purplish) produces fruits with 1–6 segments. Typically, the terminal segment contains a seed and is 7–9 mm long and 5–6 mm wide, while the other segments are sterile and ca. 2 mm long and 2 mm wide. It is found on Baltra, Daphne, Española, Floreana, Isabela (A,D), Marchena, Pinta, Pinzón, Rábida, San Cristóbal, Santa Cruz, Santiago, Seymour, and one or more of the smaller islands.

Desmodium procumbens (corolla pinkish to white) produces fruits with 3–6 sections, all of which are fertile and ca. 2.5–5 mm long and 1.5–3 mm wide. It occupies Baltra, Española, Floreana, Genovesa, Isabela (A,CA,D,SN), Marchena, Pinta, Pinzón, Rábida, San Cristóbal, Santa Cruz, Santa Fe, Santiago, Seymour, Wolf, and one or more of the smaller islands.

265. *Desmodium incanum* (tick trefoil), flowers

Scientific Name: Tephrosia decumbens Benth. (Photo 266)
Common Name: Hoary pea

Family: Fabaceae (Pea)

Range: Native; also known from Mexico and Central America.

Islands Inhabited: Baltra, Daphne, Española, Fernandina, Floreana, Isabela (A,CA, D,SN), Marchena, Pinta, Pinzón, Rábida, San Cristóbal, Santa Cruz, Santa Fe, Santiago, Seymour, Islet(s)

Habitat: Arid lowlands

Description: Low-growing herb, stems to 50 cm long, covered with light gray hairs. *Leaves* alternate, odd-pinnately compound; leaflets typically 5–9, occasionally as few as 3 or as many as 11, obovate-oblong to somewhat roundish, usually 5–20 mm long, occasionally longer, both surfaces hairy, margins entire. *Flowers* opposite the leaves, in few-flowered racemes; corolla purplish to pink or blue, flower 7–8 mm long, composed of 1 large standard petal (hairy on the outer surface), 2 lateral wing petals, and 2 lower keel petals that are somewhat fused; stamens 10, 9 of these fused into a tube, the other free. *Fruit* a legume, linear, 2–5 cm long, covered with hairs; seeds numerous, light brown.

Comments: The genus name is derived from the Greek *tephros*, "ash-colored" or "light gray," in reference to the plant's hairy covering. The specific epithet, *decumbens*, means low-growing.

266. *Tephrosia decumbens* (hoary pea), flowers and fruits

Scientific Name: Sida ciliaris L. (Photo 267)
Common Name: Sida

Family: Malvaceae (Mallow)

Range: Introduced; also known from other regions of tropical America.

Islands Inhabited: Baltra, San Cristóbal, Santa Cruz

Habitat: Coastal zone and arid lowlands

Description: Low-growing perennial herb, somewhat woody at the base. *Leaves* alternate, simple; blade oblong to obovate, 6–15 mm long, lower surface covered with long stellate hairs, margins serrate. *Flowers* axillary, solitary; corolla reddish pink, with yellow at the base, somewhat rotate with 5 lobes, lobes ca. 5 mm long; stamens numerous, united in a tubular column, yellow; style branches reddish pink. *Fruit* a schizocarp, depressed ovate, 5 mm across, with 6 sections, each with 2 short awns on top; seed 1 per section, dark brown.

Comments: The specific epithet of this tiny herb, *ciliaris,* refers to the fine hairs on the margins of the flower's sepals and the leaf's stipules.

267. *Sida ciliaris* (sida), flower and fruits

Scientific Name: Oxalis corymbosa DC. (Photo 268)
Common Name: Pink wood sorrel

Family: Oxalidaceae (Oxalis)

Range: Introduced; also known from other tropical regions throughout the world.

Islands Inhabited: San Cristóbal, Santa Cruz

Habitat: Moist uplands

Description: Low-growing perennial herb to 30 cm tall, arising from an underground bulb. *Leaves* alternate, palmately compound; leaflets 3, obcordate, 1.4–3 cm long, 2–4.5 cm wide, both surfaces with tiny reddish dots, margins fringed with fine hairs. *Flowers* in axillary, umbel-like cymes; corolla pink with dark pink veins, greenish in the center, petals 5, 1.5 cm long; stamens 10. *Fruit* a capsule; seeds numerous.

Comments: The flower color of pink wood sorrel makes it impossible to confuse this species with the other members of *Oxalis.*

268. *Oxalis corymbosa* (pink wood sorrel), flowers

Scientific Name: Polygala sancti-georgii Riley (Photos 269–70)
Common Name: St. George's milkwort

Family: Polygalaceae (Milkwort)

Range: Endemic

Islands Inhabited: Floreana, Rábida, Santa Cruz, Santiago

Habitat: Arid lowlands

Description: Annual herb to 60 cm tall, stems smooth, reddish. *Leaves* alternate, simple; blade oblanceolate to spatulate, 6–25 mm long, leathery, with a bluish, waxy covering, margins entire. *Flowers* in terminal racemes, each flower with 1 bract and 2 bracteoles beneath; calyx of 5 sepals, the outer 3 green with white margins (1.5–2 mm long), the inner 2 (wings) purplish to white (4–6.5 mm long), 5-nerved; corolla purplish to white, ca. 3–3.5 mm long, petals 3, the lower one boat-shaped (keel) with a conspicuous 12- to 16-lobed crest at its tip; stamens 8, fused together into a tubelike sheath with a lengthwise split above. *Fruit* a capsule, oblong, 3.5–4 mm long; seeds 2, hairy, covered with an extra piece of white tissue ca. one-fifth the length of the seed.

Comments: This species includes two varieties. *Polygala sancti-georgii* var. *sancti-georgii* (Photos 269–70) is known only from Floreana. *Polygala sancti-georgii* var. *oblanceolata* J. T. Howell occurs on Rábida, Santa Cruz, and Santiago. These two varieties differ primarily in that the former has leaves that are broadly spatulate, while those of the latter are narrowly to broadly oblanceolate. In addition, the crest at the tip of the keel of var. *sancti-georgii* is 14- to 16-lobed, while that of var. *oblanceolata* is 12- to 14-lobed.

Polygala galapageia Hook. f. (Galápagos milkwort) differs from *P. sancti-georgii* in producing flowers with 3-nerved wings and an inconspicuous crest at the tip of the keel. In addition, its flowers are typically white. This species includes two varieties, both rare endemics. These are var. *galapageia* and var. *insularis* (A. W. Bennett) B. L. Rob. The former is found on Floreana, Isabela (A,D), Marchena, Rábida, San Cristóbal, Santa Cruz, and Santiago, while the latter occurs on Floreana, Isabela (D), Marchena, Pinta, San Cristóbal, and Santa Cruz. One of the major differences between these two varieties is the size and shape of their racemes. Those of var. *galapageia* are usually longer, reaching a length of up to 12 cm, while those of var. *insularis* are typically not more than 4 cm long. Additionally, var. *galapageia* may reach an overall height of 1 m, whereas var. *insularis* is usually 30–45 cm tall.

Polygala anderssonii B. L. Rob. (Andersson's milkwort) differs from all of these in having stems that are slightly hairy rather than smooth and wings that are 5- to 7-nerved. This rare endemic has purplish to white flowers and is found on Isabela (A), Santa Cruz, and Santiago.

The genus derives its name from the Greek *poly*, "much," and *gala*, "milk." This comes from an ancient belief that if cows eat these plants it will enhance their milk production.

269. (above) *Polygala sancti-georgii* var. *sancti-georgii* (St. George's milkwort), flowers

270. (right) *Polygala sancti-georgii* var. *sancti-georgii* (St. George's milkwort), flowers

Scientific Name: Browallia americana L. (Photo 271)
Common Name: Bush violet

Family: Solanaceae (Potato)

Range: Introduced; also known from other regions of tropical America.

Islands Inhabited: Santa Cruz

Habitat: Moist uplands

Description: Annual herb to 1 m tall. *Leaves* alternate, simple; blade ovate, 1.5–6 cm long, margins entire. *Flowers* axillary, solitary; corolla purple to blue with 2 white and yellow spots in the center, tubular and salverform with 5 lobes, 1–1.3 cm long, 1–1.2 cm across, somewhat hairy on the outside; stamens 4. *Fruit* a capsule, brown, ovoid, 6–8 mm long; seeds numerous.

Comments: This plant frequents disturbed habitats in forested areas. The genus was named in honor of the Swedish botanist John Browall (1707–55).

271. *Browallia americana* (bush violet), flowers

See also

Ionopsis utricularioides (p. 189)
Polygonum opelousanum (p. 192)
  P. hydropiperoides (p. 192)
Portulaca grandiflora (p. 212)
Sida paniculata (p. 206)
  S. rupo (p. 206)

Herbs with Alternate Leaves and Blue Flowers

Scientific Name: Commelina diffusa Burm. f. (Photo 272)
Common Names: Dayflower, *chiriyuyo*

Family: Commelinaceae (Spiderwort)

Range: Native; also known from other regions of tropical America (Florida to Peru).

Islands Inhabited: Fernandina, Floreana, Isabela (CA,D,E,SN,W), Pinta, Pinzón, San Cristóbal, Santa Cruz, Santiago

Habitat: Moist uplands

Description: Low-growing perennial herb, stems to 80 cm long, rooting at the nodes. *Leaves* alternate, simple; petiole forming a sheath on the stem, sheath fringed with fine white hairs; blade ovate, 1.5–5 cm long, somewhat fleshy, margins entire. *Flowers* in terminal or axillary cymes, each cyme or pair of cymes with a large folded bract beneath; many of the flowers are sterile; corolla usually blue, occasionally white, petals 3, the upper two 4–5 mm long, the lower one shorter; fertile stamens 3, sterile stamens 2–3. *Fruit* a capsule, broadly ellipsoid, 4–5 mm long; seeds 1–5, black, with netlike markings.

Comments: This genus was named for two Dutch botanists, Johan (1629–92) and Caspar (1667–1731) Commelin. An interesting story holds that there was another Commelin as well—one who never contributed to the field of botany. Thus, the flowers have two larger petals that represent the two botanists, and one smaller petal that represents the third Commelin and his "lesser" accomplishments. These delicate flowers get their common name from the fact that they appear at sunrise and usually last only until midday, at which time they wither. In fact, the petals are so fragile that they appear to melt away when rubbed between the fingers. Another curious feature of this plant is that its stems are somewhat slimy and sticky to the touch when crushed. Dayflower appears to prefer somewhat shady areas, although at times it inhabits open fields. Galápagos tortoises are known to feed on this plant.

272. *Commelina diffusa* (dayflower), flower

See also

Browallia americana (p. 228)
Clitoria ternatea (p. 222)
Desmodium limense (p. 223)
Elytraria imbricata (p. 216)
Evolvulus convolvuloides (p. 221)
Heliotropium indicum (p. 219)
Plumbago coerulea (p. 190)
Pseudelephantopus spicatus (p. 218)
P. spiralis (p. 218)
Tephrosia decumbens (p. 224)

Herbs with Alternate Leaves and Green Flowers

Scientific Name: Amaranthus spinosus L. (Photo 273)
Common Name: Spiny amaranth

Family: Amaranthaceae (Amaranth)

Range: Introduced; also known from other tropical regions throughout the world.

Islands Inhabited: Isabela (CA,SN), San Cristóbal, Santa Cruz

Habitat: Arid lowlands and moist uplands

Description: Annual herb to 1 m tall, branching, stems often reddish. *Leaves* alternate, simple; blade rhombic-ovate, elliptic, or lanceolate-oblong, 4–7 cm long, typically having a pair of spines at the base, 1–2 cm long, margins entire. *Flowers* unisexual (plants monoecious), in terminal spikes and axillary clusters. Terminal spikes have staminate flowers at the top and pistillate flowers at the bottom; axillary clusters have only pistillate flowers. Each flower has 2 bracteoles beneath; calyx greenish, sepals 5, 1.5 mm long; corolla absent. Staminate flowers with 5 stamens. *Fruit* a utricle, 1.5 mm long, often opening by means of a caplike lid; seed 1, dark brown to black, shiny.

Comments: Spiny amaranth, one of nine species in the genus *Amaranthus,* is commonly observed along sidewalks and roadsides. Each of the other species has the same general appearance, but without spines, and can be seen throughout the archipelago. Three of these species are endemics.

273. *Amaranthus spinosus* (spiny amaranth), flowers

Scientific Name: Euphorbia cyathophora J. A. Murray (Photo 274)
Common Name: Mexican fire plant

Family: Euphorbiaceae (Spurge)

Range: Cultivated; also known from other tropical regions throughout the world, originally from the southeastern United States to northern South America.

Islands Inhabited: Santa Cruz

Habitat: Arid lowlands

Description: Annual herb to 50 cm tall, containing a milky sap. *Leaves* alternate near base, opposite above, simple, lobed, linear, ovate, or fiddle-shaped, to ca. 15 cm long, margins entire to dentate. *Flowers* unisexual (plants monoecious), in terminal clusters of greenish cup-shaped structures known as "cyathia;" each cluster has several green bracts with reddish bases beneath. Each cyathium has a single, yellowish green, 2-lipped, flattened gland. Staminate flowers numerous in each cyathium; calyx absent; corolla absent; stamen 1. Pistillate flowers 1 per cyathium, projecting from the mouth of the cyathium, greenish; calyx absent; corolla absent. *Fruit* a capsule, broadly ovoid, 3–4 mm long; seed 1.

Comments: Mexican fire plant is a common ornamental, and because of the bright red and green bracts located near its flowers, one should have no trouble making a correct identification. Its flowers are visited by the Galápagos carpenter bee (*Xylocopa darwini*) throughout the day. The genus was named for the first-century physician Euphorbus, who cared for the king of Mauretania.

Another cultivated member of this genus is *Euphorbia pulcherrima* Willd. ex Klotzsch (poinsettia, *flor de pascuas*). It too is known from other tropical regions throughout the world, but it is thought to have originated in Mexico. It inhabits San Cristóbal and Santa Cruz and may be found in both the arid lowlands and moist uplands. This shrub has flowers that are similar to those of *E. cyathophora*. However, the bracts beneath each cluster of cyathia are completely red and are often mistaken for petals. This beautiful symbol of Christmas, planted as a yard ornamental, is actually quite toxic. Eating any of its parts can cause vomiting, diarrhea, shock, and even death. In addition, its milky sap can irritate the skin.

274. *Euphorbia cyathophora* (Mexican fire plant), flowers

Scientific Name: Phyllanthus caroliniensis Walt. (Photos 275–76)
Common Name: Phyllanthus

Family: Euphorbiaceae (Spurge)

Range: Native; also known from other regions of tropical America (United States to Argentina).

Islands Inhabited: Fernandina, Floreana, Isabela (A,CA,D,SN), Pinta, San Cristóbal, Santa Cruz, Santiago

Habitat: Moist uplands

Description: Annual or short-lived perennial herb to 30 cm tall. *Leaves* alternate, simple; blade elliptic to obovate, 3–15 mm long, margins entire. *Flowers* unisexual (plants monoecious), in axillary clusters of 1–2 staminate and 1–2 pistillate flowers. Staminate flower calyx yellowish green, sepals 6, less than 1 mm long; corolla absent; stamens 3. Pistillate flower calyx yellowish green or sometimes tinged with red, sepals 6, 0.7–1.4 mm long; corolla absent. *Fruit* a capsule, ca. 1.5–2 mm across; seeds 3–6.

Comments: All Galápagos members of this species represent subsp. *caroleniensis.* It appears to prefer open, grassy areas such as those found in the highlands. Relatively nonshowy, this small herb is often overlooked.

Phyllanthus acidus (L.) Skeels (gooseberry tree) is also known from Santa Cruz. However, it is a tree and easily distinguishable from *P. caroliniensis.*

275. (above) *Phyllanthus caroliniensis* subsp. *caroleniensis* (phyllanthus), flower

276. (left) *Phyllanthus caroliniensis* subsp. *caroleniensis* (phyllanthus), fruits

Scientific Name: Epidendrum spicatum Hook. f. (Photo 277)
Common Name: Buttonhole orchid

Family: Orchidaceae (Orchid)

Range: Endemic

Islands Inhabited: Floreana, Isabela (A,SN), Pinta, San Cristóbal, Santa Cruz, Santiago

Habitat: Moist uplands

Description: Perennial herb to 50 cm tall, growing as an epiphyte on trees. *Leaves* alternate, simple, lanceolate or ovate-lanceolate, to 15 cm long, fleshy, sheathing at base, margins entire. *Flowers* in terminal racemes, each flower with a greenish bract beneath; corolla yellowish green to whitish green, petals 3; stamen 1. The two lateral petals are yellowish green, linear, and to 11 mm long. The third petal, referred to as the "lip" or "labellum," is attached to a central structure known as the "column" or "gynandrium," which is a compound structure composed of the stamen, style, and stigma. The labellum itself is whitish green, 3-lobed with jagged margins, and 10 mm across. *Fruit* a capsule, brown, ellipsoid, to ca. 2.5 cm long; seeds numerous, dust-like.

Comments: The genus name is derived from the Greek *epi*, "upon," and *dendron*, "a tree." This name refers to the orchid's epiphytic nature.

277. *Epidendrum spicatum* (buttonhole orchid), flowers

Scientific Name: Pothomorphe peltata (L.) Miq. (Photo 278)
Common Name: Pothomorphe

Family: Piperaceae (Pepper)

Range: Cultivated escape; also known from other regions of tropical America.

Islands Inhabited: Isabela (SN), Santa Cruz

Habitat: Moist uplands

Description: Perennial herb to 2 m tall, stems covered with minute yellowish brown glands. *Leaves* alternate, simple; blade roundish cordate, to ca. 18 cm long, covered with tiny yellowish brown glands, margins entire. *Flowers* in axillary spikes 5–10 cm long, each flower with a fringed, greenish, triangular bract beneath; calyx absent; corolla absent; stamens 2. *Fruit* drupelike, 3-angled; seed 1.

Comments: Pothomorphe peltata tends to grow in disturbed areas such as roadsides and fields. The specific epithet, *peltata,* refers to the fact that the leaf is attached to its stalk in a peltate fashion (Hamann 1974a). In other words, the stalk is attached to the central portion of the blade's lower surface rather than at the base.

278. *Pothomorphe peltata* (pothomorphe), flowers

See also

Habenaria alata (p. 188)
H. distans (p. 188)
Plantago galapagensis (p. 293)
P. major (p. 293)
Polygonum punctatum (p. 192)
Porophyllum ruderale (p. 217)

Herbs with Alternate Leaves and Brown Flowers

Scientific Name: Cyperus anderssonii Boeck. (Photo 279)
Common Name: Andersson's sedge

Family: Cyperaceae (Sedge)

Range: Endemic

Islands Inhabited: Baltra, Darwin, Fernandina, Floreana, Genovesa, Isabela (A,CA,D, SN), Marchena, Pinta, Pinzón, Rábida, San Cristóbal, Santa Cruz, Santa Fe, Santiago, Wolf, Islet(s)

Habitat: Coastal zone, arid lowlands, and moist uplands

Description: Perennial herb with a single upright stem to 70 cm tall, or several stems clumped together. *Leaves* alternate and basal, simple; blade linear, to more than 70 cm long, 2–6 mm wide, the lower portion purplish brown and completely surrounding the stem, margins entire. *Flowers* in terminal, umbel-like corymb composed of stalked spikes to 9 cm long; 3–7 leaflike bracts beneath, to 30 cm long; spikes to ca. 2 cm long and 7 mm broad, composed of tight clusters of numerous spikelets; spikelets 2- to 3-flowered, 3 mm long; each flower with a bract beneath, yellowish brown to reddish brown; calyx absent; corolla absent; stamens 3. *Fruit* an achene, dark brown, elliptical, ca. 1 mm long, somewhat 3-sided; seed 1.

Comments: This is one of 18 species of *Cyperus* found in the Galápagos. Three of these are endemic, 11 native, and 4 introduced. Members of the genus may be found in a variety of habitats, ranging from near the shoreline all the way to the highlands. Some appear to prefer sandy soils, while others thrive in wetlands or meadows. The different species are difficult to distinguish, especially for the novice. In fact, these plants are often mistaken for grasses. However, knowledge of a few simple characteristics will prevent this mistake. For example, sedges typically have triangular stems, while grasses have rounded stems; sedges have solid stems, while grasses have hollow ones; and the leaf bases of sedges completely surround and are directly attached to the stem, while the leaf bases of grasses only partially surround the stem and may be pulled away from it.

279. *Cyperus anderssonii* (Andersson's sedge)

Scientific Name: Cyperus ligularis L. (Photo 280)
Common Name: Sedge

Family: Cyperaceae (Sedge)

Range: Native; also known from other tropical regions throughout the world.

Islands Inhabited: Fernandina, Isabela (A,CA,D,SN), Marchena, Pinta, Santa Cruz

Habitat: Coastal zone

Description: Perennial herb with a single upright stem to 80 cm tall. *Leaves* alternate and basal, simple; blade linear, to more than 80 cm long, 8–12 mm wide, the lower portion reddish brown to purplish brown and completely surrounding the stem, margins rough to minutely serrate. *Flowers* in terminal, umbel-like corymb composed of stalked spikes to 10 cm long; 5–8 leaflike bracts beneath, to ca. 65 cm long; spikes to ca. 2.5 cm long and 1.2 cm broad, composed of tight clusters of numerous spikelets; spikelets 2- to 6-flowered, 4–6 mm long; each flower with a bract beneath, brown to reddish brown; calyx absent; corolla absent; stamens 3. *Fruit* an achene, brown, obovate, 1.5 mm long, somewhat 3-sided; seed 1.

Comments: This species is similar to *Cyperus anderssonii* (Andersson's sedge). It differs in having wider leaves, wider spikes, and longer spikelets.

280. *Cyperus ligularis* (sedge)

Scientific Name: Aristida subspicata Trin. & Rupr. (Photo 281)
Common Name: Galápagos three-awn grass

Family: Poaceae (Grass)

Range: Endemic

Islands Inhabited: Baltra, Daphne, Fernandina, Floreana, Genovesa, Isabela (A,D), Marchena, San Cristóbal, Santa Cruz, Santa Fe, Santiago, Seymour, Islet(s)

Habitat: Arid lowlands

Description: Perennial herb typically to 50 cm tall, occasionally taller, stems growing in clumps. *Leaves* alternate, simple; blade narrow, to 20 cm long, with a basal sheath surrounding the stem and an appendage (ligule) where the leaf blade and sheath meet, both surfaces somewhat rough, margins rough. *Flowers* in spikelets on terminal, spikelike panicles 6–10 cm long, yellowish brown, surfaces rough. Each spikelet possesses a single floret with 2 rough bracts (glumes) beneath. The floret consists of a single flower enclosed in 2 additional, unequal bracts. The larger (lemma) is 8–10 mm long, with 3 awns, the longest being 10–15 mm. The smaller bract is called the "palea." Calyx absent; corolla absent; stamens 3. *Fruit* a grain; seed 1.

Comments: Aristida villosa B. L. Rob. & Greenm., a rare endemic that occurs on Baltra, Floreana, Pinzón, Rábida, Santa Fe, Santiago, and Seymour, also possesses spikelike panicles. It differs from *A. subspicata* in that its sheaths are covered with long, white hairs.

Aristida divulsa Andersson and *A. repens* Trin. are endemics that have non-spikelike inflorescences. In other words, their spikelets are more loosely arranged. They differ from each other in that *A. divulsa* is an erect perennial, whereas *A. repens* is an annual that tends to lean toward the ground as it grows. *Aristida divulsa* is considered rare but is known from Fernandina, Isabela (A,CA,D), Marchena, Pinta, San Cristóbal, Santa Cruz, and Santiago. *Aristida repens* is found on Fernandina, Floreana, Genovesa, Isabela (A,D,SN,W), Marchena, Pinta, San Cristóbal, Santa Cruz, Santa Fe, Santiago, and one or more of the smaller islands.

281. *Aristida subspicata* (Galápagos three-awn grass)

Scientific Name: Chloris virgata Sw. (Photo 282)
Common Name: Feather fingergrass

Family: Poaceae (Grass)

Range: Native; also known from other tropical regions throughout the world.

Islands Inhabited: Baltra, Daphne, Española, Floreana, San Cristóbal, Santa Cruz, Santa Fe, Seymour, Islet(s)

Habitat: Arid lowlands

Description: Annual herb to ca. 1 m tall, stems round, smooth, usually growing in clumps. *Leaves* alternate, simple; blade narrow, to 25 cm long, with a basal sheath surrounding the stem and an appendage (ligule) where the leaf blade and sheath meet, both surfaces smooth to somewhat rough, margins rough. *Flowers* in spikelets on terminal, feathery, fingerlike racemes usually 5–6 cm long, yellowish brown. Each spikelet is 3–3.5 mm long, has 2 bracts (glumes) beneath, and possesses a single fertile floret and 1 or more undeveloped florets. The floret consists of a flower enclosed in 2 additional, unequal bracts; the larger is called the "lemma," while the smaller is known as the "palea." The lemma of the fertile floret is widest above the middle. It has fine hairs to 3 mm long on its upper margins, a crown of hairs at its apex, and an awn 6–10 mm long. Calyx absent; corolla absent; stamens 3. *Fruit* a grain; seed 1.

Comments: Chloris mollis (Nees) Swallen and *C. radiata* (L.) Sw. differ from *C. virgata* in having fertile floret lemmas that are widest at or below the middle. They differ from each other in that *C. mollis* has round stems, while *C. radiata* has flattened stems. In addition, the fertile florets of the former are purplish, while those of the latter are yellowish to brownish with only occasional touches of purple. *Chloris mollis* inhabits Española and Floreana, whereas *C. radiata* is known from Española, Floreana, Isabela (SN), Santa Cruz, and Santiago. It's possible that a fourth species, *C. pycnothrix* Trin., also inhabits the archipelago. Specimens identified as this plant have been collected on Santa Cruz. However, according to Anderson (1974), the only substantial difference between these collections and *C. radiata* is that the former have longer lemma awns that fall within the range typical of *C. pycnothrix* (10–45 mm). Those of *C. radiata* usually do not exceed 13 mm.

This genus illustrates how opinions may differ on the circumstances surrounding the arrival of a particular species in the archipelago. Porter (1983) lists all four species as introduced weeds, while Lawesson et al. (1987) consider them natives.

282. *Chloris virgata* (feather fingergrass)

Scientific Name: Eragrostis ciliaris (L.) R. Br. (Photo 283)
Common Name: Eragrostis

Family: Poaceae (Grass)

Range: Native; also known from other tropical regions throughout the world.

Islands Inhabited: Baltra, Española, Fernandina, Floreana, Genovesa, Isabela (A,CA,D, SN,W), Marchena, Pinta, Pinzón, Rábida, San Cristóbal, Santa Cruz, Santiago

Habitat: Arid lowlands

Description: Annual herb to ca. 40 cm tall, stems slender. *Leaves* alternate, simple; blade narrow, to 10 cm long, with a basal sheath surrounding the stem and an appendage (ligule) where the leaf blade and sheath meet, both surfaces smooth to somewhat hairy, ligule margin with fine hairs, leaf margins smooth. *Flowers* in somewhat flattened spikelets on terminal, spikelike panicles ca. 5–10 cm long, yellowish brown. Each spikelet, 2–3 mm long, possesses 6–12 florets, with 2 bracts (glumes) beneath. The floret consists of a flower enclosed in 2 additional, unequal bracts. The larger bract (lemma) is 1–1.5 mm long, while the smaller bract (palea) is slightly shorter and has a fringe of fine hairs. Calyx absent; corolla absent; stamens 3. *Fruit* a grain; seed 1.

Comments: Eragrostis cilianensis (All.) Lut. and *E. mexicana* (Hornem.) Link differ from *E. ciliaris* in having panicles that are open rather than spikelike. They differ from each other in that *E. cilianensis* has spikelets approximately 2.5 mm wide, while those of *E. mexicana* are about 1.5 mm wide. *Eragrostis cilianensis* is a native found on Baltra, Daphne, Darwin, Española, Floreana, Genovesa, Isabela (SN), Pinzón, San Cristóbal, Santa Cruz, Santa Fe, Santiago, Wolf, and one or more of the smaller islands. *Eragrostis mexicana*, also a native, occurs on Isabela (CA,D,SN,W), Marchena, and Pinta.

Eragrostis pilosa (L.) Beauv. (India lovegrass) is an introduced weed. It is known only from Isabela (SN) and Santiago. The genus name is based on the Greek *Eros*, the god of love, and *agrostis*, a kind of grass.

283. *Eragrostis ciliaris* (eragrostis)

Scientific Name: Pennisetum purpureum Schum. (Photo 284)
Common Names: Elephant grass, Napier grass, *pasto elefante*

Family: Poaceae (Grass)

Range: Cultivated escape; also known from other tropical regions throughout the world, originally from Africa.

Islands Inhabited: Floreana, Isabela (SN), San Cristóbal, Santa Cruz

Habitat: Arid lowlands and moist uplands

Description: Perennial herb 2–6 m tall, stems typically smooth, to 2 cm in diameter, usually growing in large clumps. *Leaves* alternate, simple; blade narrow, 30–120 cm long, with a basal sheath surrounding the stem and an appendage (ligule) where the leaf blade and sheath meet, upper surface somewhat hairy, margins rough. *Flowers* in spikelets (solitary or in clusters of 2–5 and surrounded by bristles) on terminal, dense, spikelike panicles 8–30 cm long, yellowish brown to purplish. Each spikelet, ca. 5 mm long, possesses 1 sterile staminate floret and 1 fertile floret, with two bracts (glumes) beneath. The floret consists of a flower enclosed in 2 additional, unequal bracts. The larger is known as the "lemma," while the smaller is called the "palea." Calyx absent; corolla absent; stamens 3. *Fruit* a grain; seed 1.

Comments: The genus derives its name from the Latin *penna*, "feather," and *seta*, "bristle." This refers to the feathery bristles surrounding each spikelet. The specific epithet alludes to the panicles, which are sometimes purplish. This plant is often used as forage.

Another member of the grass family that has escaped cultivation is *Bambusa guadua* HBK. (bamboo) (Photo 285). It can be seen along the roadsides of Isabela (SN), San Cristóbal, and Santa Cruz.

Pennisetum pauperum Steud. (Galápagos elephant grass), an endemic, differs from *P. purpureum* in having slenderer stems. In addition, its panicles are slenderer and possess few bristles. It is found on Fernandina, Isabela (D,E,W), and Santiago.

284. *Pennisetum purpureum* (elephant grass)

285. *Bambusa guadua* (bamboo)

Scientific Name: Sporobolus virginicus (L.) Kunth (Photo 286)
Common Names: Beach dropseed, *hierba de orilla, pasto de playa*

Family: Poaceae (Grass)

Range: Native; also known from other tropical regions throughout the world.

Islands Inhabited: Española, Fernandina, Floreana, Genovesa, Isabela (A,CA,SN), San Cristóbal, Santa Cruz, Santiago, Islet(s)

Habitat: Coastal zone

Description: Perennial herb 10–40 cm tall, stems somewhat wiry, much-branched, widely spreading or creeping. *Leaves* alternate, in 2 vertical rows on opposite sides of the stem, simple; blade narrow, typically 3–8 cm long, occasionally longer, with a basal sheath surrounding the stem and an appendage (ligule) where the leaf blade and sheath meet, upper surface smooth or with a few long hairs near the base, margins entire. *Flowers* in spikelets on terminal, dense, spikelike panicles 2–10 cm long, yellowish brown to grayish. Each spikelet, 2–2.5 mm long, possesses a single floret with 2 bracts (glumes) beneath. The floret consists of a flower enclosed in 2 additional, unequal bracts. The larger is known as a "lemma," while the smaller is called a "palea." Calyx absent; corolla absent; stamens 3. *Fruit* a grain; seed 1.

Comments: This grass frequents sandy, coastal areas. *Sporobolus pyramidatus* (Lam.) Hitchc., also native, occurs on Española, Fernandina, and Isabela (CA,D). It differs from *S. virginicus* in that it is not a creeping plant and its leaves are not obviously arranged in 2 vertical rows on opposite sides of the stem. *Sporobolus indicus* (L.) R. Br. (West Indian dropseed) also differs from *S. virginicus* in these two characters. In addition, its panicles are longer (usually 20–40 cm). It is a native that inhabits Fernandina, Floreana, Isabela (A,CA,SN), San Cristóbal, and Santa Cruz.

286. *Sporobolus virginicus* (beach dropseed)

Herbs with Opposite Leaves and White Flowers

Scientific Name: Sesuvium edmonstonei Hook. f. (Photos 287–88)
Common Name: Galápagos carpetweed

Family: Aizoaceae (Carpetweed)

Range: Endemic

Islands Inhabited: Darwin, Española, Fernandina, Floreana, Genovesa, Isabela (SN), Rábida, San Cristóbal, Santa Cruz, Santiago, Seymour, Wolf, Islet(s)

Habitat: Coastal zone

Description: Low-growing perennial herb, sometimes with a woody base, stems fleshy, covered with scales. *Leaves* opposite, simple, oblanceolate, typically to 3 cm long, fleshy. *Flowers* axillary, solitary; calyx white, 5-lobed, lobes to ca. 6 mm long; corolla absent; stamens numerous. *Fruit* a capsule, ovoid with a caplike lid; seeds numerous, black, with wrinkled seed coats.

Comments: Sesuvium edmonstonei often forms dense carpets over rocky and sandy terrain, thus the common name. During the cool season this plant takes on a spectacular bright orange-red color. The specific epithet honors Thomas Edmonston (1825–46), a naturalist who visited the archipelago in January 1846 and collected several plant specimens. Later that same month, while visiting mainland Ecuador, he died tragically from an accidental gunshot wound (Porter 1980a).

287. (top) *Sesuvium edmonstonei* (Galápagos carpetweed)

288. (right) *Sesuvium edmonstonei* (Galápagos carpetweed), flowers

Scientific Name: Trianthema portulacastrum L. (Photo 289)
Common Name: Trianthema

Family: Aizoaceae (Carpetweed)

Range: Native; also known from other tropical regions throughout the world.

Islands Inhabited: Baltra, Española, Floreana, Genovesa, Isabela (SN), Pinzón, Rábida, San Cristóbal, Santa Cruz, Santa Fe, Seymour, Islet(s)

Habitat: Coastal zone

Description: Low-growing herb, stems somewhat fleshy, red. *Leaves* opposite, 1 of each pair larger in size, bases meeting at the stem to form a tiny cup, simple; blade typically obovate or spatulate, 1–2 cm long, margins entire. *Flowers* axillary, solitary; calyx white or purplish, 5-lobed, lobes 5 mm long; corolla absent; stamens numerous. *Fruit* a capsule, ovoid with a caplike lid; seeds 1–2, dull black, reniform.

Comments: The genus name comes from the Greek *tri*, "three," and *anthemon*, "flower," because certain other members of the genus produce 3 flowers in each inflorescence.

289. *Trianthema portulacastrum* (trianthema), flowers

Scientific Name: Adenostemma platyphyllum Cass. (Photo 290)
Common Name: Adenostemma

Family: Asteraceae (Sunflower)

Range: Introduced; also known from other tropical regions throughout the world.

Islands Inhabited: Santa Cruz

Habitat: Moist uplands

Description: Annual herb to 1 m tall. *Leaves* usually opposite, upper ones sometimes alternate, simple; blade ovate or elliptic to variously shaped, 5–15 cm long, margins coarsely serrate. *Flowers* in discoid heads, disc 5–8 mm across; heads numerous; flowers white, perfect, style split into 2 white club-shaped branches that are much longer than the corolla. *Fruit* an achene, brown, to 3 mm long, warty, with 3 glandular awns on top, 1 mm long; seed 1.

Comments: This plant, referred to as *Adenostemma lavenia* (L.) Kuntze in Wiggins and Porter (1971), is relatively uncommon. The genus name appears to have come from the Greek *aden*, "gland," and *stemma*, "crown," which refer to the 3 glandular awns atop each fruit. Interestingly, these glandular awns are extremely sticky and provide an ideal means for dispersal by adhering to animals that brush against them. The fruits often cling to the clothing of humans as well, which may account for its introduction to the archipelago.

290. *Adenostemma platyphyllum* (adenostemma), flowers

Scientific Name: Blainvillea dichotoma (Murr.) A. Stewart (Photo 291)
Common Name: Blainvillea

Family: Asteraceae (Sunflower)

Range: Native; also known from other tropical regions throughout the world.

Islands Inhabited: Baltra, Daphne, Española, Fernandina, Floreana, Isabela (A,CA,D, SN), Pinta, Pinzón, Rábida, San Cristóbal, Santa Cruz, Santa Fe, Santiago, Seymour, Islet(s)

Habitat: Arid lowlands and moist uplands

Description: Annual herb to 1 m tall, much-branched, hairy. *Leaves* opposite or alternate (on the upper branches), simple, lanceolate to broadly ovate, usually 1.5–8 cm long, margins variously serrate. *Flowers* in radiate heads, disc 5–8 mm across; heads solitary, terminal or axillary; disc flowers white, perfect; ray flowers white, pistillate. *Fruit* an achene, black, 2–3.5 mm long, with 3 awns on top, to 1.5 mm long; seed 1.

Comments: Nice examples of this plant may be found along the trail on Santa Fe.

291. *Blainvillea dichotoma* (blainvillea), flowers

Scientific Name: Eclipta alba (L.) Hassk. (Photo 292)
Common Name: False daisy

Family: Asteraceae (Sunflower)

Range: Native; also known from other tropical regions throughout the world, originally from tropical America.

Islands Inhabited: Española, Fernandina, Floreana, Isabela (SN), San Cristóbal, Santa Cruz, Santiago

Habitat: Moist uplands

Description: Annual herb to 1 m tall, growing upward or low-growing with stem tips curving upward, stems with flattened straight hairs, often rooting at the nodes. *Leaves* opposite, simple, lanceolate, elliptic-lanceolate, or linear-lanceolate, 2–10 cm long, margins minutely serrate. *Flowers* in radiate heads ca. 6 mm across; heads generally solitary or in groups of 2–3, terminal or axillary; disc flowers white, perfect; ray flowers white, pistillate or sometimes sterile. *Fruit* an achene, 2–2.5 mm long, somewhat warty; seed 1.

Comments: False daisy is often observed along roadsides and in gardens. The genus name comes from a Greek word meaning "deficient," in reference to the absence of bristles and awns on the fruits. The specific epithet *alba* means "white" and refers to the color of the flowers.

292. *Eclipta alba* (false daisy), flowers

Scientific Name: Drymaria cordata (L.) Willd. ex Roem. & Schult. (Photo 293)
Common Name: Drymaria

Family: Caryophyllaceae (Pink)

Range: Native; also known from other tropical regions throughout the world.

Islands Inhabited: Floreana, Isabela (A,SN), San Cristóbal, Santa Cruz, Santiago

Habitat: Moist uplands

Description: Low-growing annual herb. *Leaves* opposite, simple, reniform to roundish, 8–20 mm long, 10–20 mm wide, margins entire. *Flowers* in terminal or axillary cymes; corolla white, petals 5, 2–2.5 mm long, each petal deeply 2-lobed; stamens 2–3. *Fruit* a capsule; seeds numerous, reddish brown, covered with tiny bumps.

Comments: Drymaria rotundifolia A. Gray, also a native, is found only on Fernandina and Isabela (A,CA,D,W). It differs from *D. cordata* in that it grows to 20 cm tall. In addition, it has leaves that are broadly ovate to triangular and are much smaller (3–7 mm long, 3–6 mm wide).

Drymaria monticola J. T. Howell (Galápagos drymaria), considered endemic and rare, is known only from Santa Cruz. It differs from *D. cordata* in that it has 4–5 stamens and its leaves are broadly ovate to almost cordate. It differs from *D. rotundifolia* in that it is low-growing and has larger leaves (10–25 mm long, 8–25 mm wide).

293. *Drymaria cordata* (drymaria), flowers

Scientific Name: Chamaesyce hirta (L.) Millsp. (Photo 294)
Common Name: Chamaesyce

Family: Euphorbiaceae (Spurge)

Range: Introduced; also known from other tropical regions throughout the world.

Islands Inhabited: Floreana, San Cristóbal, Santiago

Habitat: Arid lowlands

Description: Low-growing annual herb to 30 cm long, containing a milky sap, young stems densely covered with hairs. *Leaves* opposite, simple; blade ovate to lanceolate or rhombic, to 3.5 cm long, margins serrate. *Flowers* unisexual (plants monoecious), in dense terminal and axillary clusters of reddish, cup-shaped structures known as "cyathia." Each cyathium has 4 tiny, purple glands that often possess white appendages resembling petals. Staminate flowers 6–12 per cyathium; calyx absent; corolla absent; stamen 1. Pistillate flowers 1 per cyathium, projecting from the mouth of the cyathium, greenish; calyx absent; corolla absent. *Fruit* a capsule, ovoid, to 1 mm long, covered with hairs; seed 1, 4-angled.

Comments: This is one of 11 species of *Chamaesyce* occurring in the archipelago. Its distribution, near towns, roads, and trails, is typical of introduced organisms. Although small, the clusters of cyathia characteristic of this plant are actually quite attractive. Several members of this genus are somewhat similar in appearance, and their accurate identification requires scrutiny.

294. *Chamaesyce hirta* (chamaesyce), flowers

Scientific Name: Hyptis rhomboidea Mart. & Gal. (Photo 295)
Common Name: Hyptis

Family: Lamiaceae (Mint)

Range: Introduced; also known from other tropical regions throughout the world.

Islands Inhabited: Isabela (A,SN), San Cristóbal, Santa Cruz, Santiago

Habitat: Moist uplands

Description: Perennial herb to 2 m tall, stems square in cross section, becoming reddish, often rooting at the nodes. *Leaves* opposite, simple; petiole somewhat winged; blade rhombic-ovate, 3–12 cm long, margins serrate. *Flowers* in axillary, stalked, dense heads; corolla white with purplish markings, tubular with 2 distinct lips, ca. 4 mm long, upper lip notched, lower lip 3-lobed and somewhat saclike; stamens 4, resting on the lower lip. *Fruit* consisting of 4 nutlets, ovoid, ca. 1 mm long; seeds 1 per nutlet.

Comments: Two other members of the genus also have their flowers arranged in heads. These are *H. gymnocaulos* Epling, a rare endemic known only from Isabela (A), and *H. sidaefolia* (L'Hér.) Briq., an introduced species, found only on Santa Cruz. They differ from each other in that the former is somewhat woody to shrubby, whereas the latter is an herb. *Hyptis sidaefolia* differs from *H. rhombifolia* in having leaf bases that are rounded to almost cordate, while the latter has leaf bases that are wedge-shaped.

The final two species have their flowers arranged in whorls on a long spike. *Hyptis spicigera* Lam., a native known from Fernandina, Floreana, Isabela (D,SN), and Santiago, is herbaceous. *Hyptis mutabilis* (A. Rich.) Briq., an introduced species known only from Isabela (SN), is somewhat woody.

295. *Hyptis rhomboidea* (hyptis), flowers

Scientific Name: Cuphea racemosa (L. f.) Spreng. (Photo 296)
Common Name: White Cuphea

Family: Lythraceae (Loosestrife)

Range: Introduced; also known from other regions of tropical America.

Islands Inhabited: San Cristóbal, Santa Cruz

Habitat: Moist uplands

Description: Annual herb to ca. 50 cm tall, becoming somewhat woody at the base, stems reddish, with glandular hairs. *Leaves* opposite, simple, obovate, elliptic, or oblong, to ca. 6 cm long, lower surface with some glandular hairs, margins with fine white hairs. *Flowers* in terminal racemes; floral tube with a distinct hump at the base and dark red longitudinal stripes, covered with glandular hairs; corolla white with a pink stripe at the base of each petal, petals 6, ca. 1.5 mm long; stamens 11. *Fruit* a capsule, ovoid, ca. 4 mm long, seeds 4.

Comments: The genus name is derived from the Greek *kyphos*, "hump," in reference to the base of the floral tube. *Cuphea carthagenensis* (Jacq.) Macbr. (purple cuphea), also introduced, is found on Isabela (A,SN), San Cristóbal, and Santa Cruz. It differs from *C. racemosa* in having purple, axillary flowers (van der Werff 1977).

296. *Cuphea racemosa* (white cuphea), flowers

Scientific Name: Diodia radula (Roem. & Schult.) Cham. & Schlecht. (Photo 297)

Common Name: Buttonweed

Family: Rubiaceae (Madder)

Range: Introduced; also known from western South America.

Islands Inhabited: Fernandina, Floreana, Isabela (W), San Cristóbal, Santa Cruz, Santiago

Habitat: Moist uplands

Description: Low-growing perennial herb, stems 4-angled, to 60 cm long, somewhat woody at the base, young stems turning upward. *Leaves* opposite, simple, broadly elliptic to ovate-elliptic, 2–4.5 cm long, both surfaces covered with hairs, although more on the lower surface, margins entire. The basal portion of each of the opposite leaves is fused with the stem to form a sheath with bristles on its margins. *Flowers* in dense axillary clusters; corolla white, funnelform with 4 lobes, 3.5–6 mm long, lobes 2–3 mm long, throat densely covered with white hairs; stamens 4. *Fruit* a capsule, 2–3 mm long; seeds 2, dark brown.

Comments: Diodia radula is common in areas disturbed by agriculture, roads, and trails. In fact, the genus name comes from the Greek *diodos*, "highway." This herb can easily be mistaken for *Borreria laevis* (Lam.) Griseb. (smooth borreria), another introduced member of the Rubiaceae. However, there are a few distinguishing characteristics. For example, the corolla of *B. laevis* is only 2–2.6 mm long. If no flowers are present, then the leaves must be examined. All of the lateral veins on a leaf of *B. laevis* are attached to the midrib, while in *D. radula* the lowest 1 or 2 pairs of lateral veins are not attached to the midrib. Additionally, the leaves of *D. radula* are typically much hairier than those of *B. laevis. Borreria laevis* inhabits Fernandina, Floreana, Isabela (CA,SN), San Cristóbal, Santa Cruz, and Santiago.

297. *Diodia radula* (buttonweed), flowers

Scientific Name: Scoparia dulcis L. (Photo 298)
Common Name: Sweet-broom

Family: Scrophulariaceae (Figwort)

Range: Native; also known from other tropical regions throughout the world, originally from tropical America.

Islands Inhabited: Fernandina, Floreana, Isabela (A,SN), San Cristóbal, Santa Cruz, Santiago

Habitat: Moist uplands

Description: Perennial herb to 1.5 m tall, woody at the base, young stems with tiny glandular dots. *Leaves* opposite, simple; blade elliptic to lanceolate, 7–25 mm long, both surfaces glandular-dotted, margins serrate. *Flowers* in axillary clusters of 1–3; corolla white to pale pink, rotate with 5 lobes, 3–4 mm across, lobes ca. 2 mm long, densely hairy within; stamens 4. *Fruit* a capsule, ovoid, 3–3.5 mm long; seeds numerous.

Comments: This herb typically inhabits disturbed sites such as roadsides and old fields. Wiggins and Porter (1971) mention that it is said to repel insects.

298. *Scoparia dulcis* (sweet-broom), flower

Scientific Name: Lippia strigulosa Mart. & Gal. (Photo 299)
Common Name: Purple lippia

Family: Verbenaceae (Vervain)

Range: Native; also known from other regions of tropical America.

Islands Inhabited: Española, Isabela (SN), Pinzón, San Cristóbal, Santa Cruz

Habitat: Arid lowlands and moist uplands

Description: Low-growing, trailing, perennial herb, often purplish toward the base. *Leaves* opposite, simple; blade ovate or ovate-elliptic, 1.5–4 cm long, widest below the middle, both surfaces somewhat hairy, margins dentate and often purplish. *Flowers* in axillary heads, each flower with a purplish bracteole beneath; corolla typically white with a yellow throat, occasionally tinged with pale purple, salverform with 4 lobes, ca. 3 mm long, somewhat 2-lipped; stamens 4. *Fruit* drupaceous, ovoid, separating into 2 sections when mature; seed 1 per section.

Comments: Lippia strigulosa is referred to as *Phyla strigulosa* (Mart. & Gal.) Moldenke in Wiggins and Porter (1971). A similar species, *L. reptans* (Spreng.) HBK. (creeping lippia), is listed as *P. nodiflora* var. *reptans* (HBK.) Moldenke in Wiggins and Porter (1971). It is an introduced weed found on Floreana, San Cristóbal, and Santa Cruz, and it differs from *L. strigulosa* in that its leaves are variously shaped (obovate to spatulate) and widest at or above the middle.

299. *Lippia strigulosa* (purple lippia), flowers

Scientific Name: Priva lappulacea (L.) Pers. (Photo 300)
Common Name: Priva

Family: Verbenaceae (Vervain)

Range: Introduced; also known from other tropical regions throughout the world.

Islands Inhabited: Floreana, Isabela (A,D,SN), San Cristóbal, Santa Cruz

Habitat: Arid lowlands and moist uplands

Description: Annual or perennial herb to 1 m tall, somewhat declining with the stems turned upward, often purplish toward the base, stems 4-angled, with hooked hairs. *Leaves* opposite, simple; blade ovate, to ca. 14 cm long, both surfaces hairy, margins serrate. *Flowers* in terminal racemes to 21 cm long, each flower with a bracteole beneath; corolla white, bluish, pinkish, or purplish, salverform with 5 lobes, ca. 3.5 mm long, 2-lipped, the upper lip 2-lobed, the lower lip 3-lobed, lobes ca. 1–2 mm long; stamens 4. *Fruit* a schizocarp, oblong, ca. 3–4 mm long, separating into 2 sections when mature, enclosed in an inflated calyx covered with tiny hooked hairs; seeds 1–2 per section.

Comments: This plant is typically found in disturbed areas such as roadsides and fields. It is occasionally confused with *Teucrium vesicarium* (germander) (see Photo 318), which also possesses squarish stems, similar leaves, and an inflated calyx around each of its fruits. However, *T. vesicarium* has larger flowers (9–13 mm long), and its calyx hairs are not hooked.

300. *Priva lappulacea* (priva), flower and fruits

See also

Ageratum conyzoides (p. 271)
Bidens pilosa (p. 259)
Blechum pyramidatum (p. 267)
Boerhaavia erecta (p. 274)
Catharanthus roseus (p. 270)
Justicia galapagana (p. 268)
Mirabilis jalapa (p. 299)
Teucrium vesicarium (p. 273)

Herbs With Opposite Leaves and Yellow or Orange Flowers

Scientific Name: Asclepias curassavica L. (Photo 301)
Common Name: Butterfly weed

Family: Asclepiadaceae (Milkweed)

Range: Cultivated escape; also known from other regions of tropical America (southern United States to northern South America).

Islands Inhabited: Floreana, San Cristóbal, Santa Cruz

Habitat: Moist uplands

Description: Perennial herb to 2 m tall, occasionally somewhat woody at the base. *Leaves* opposite, simple, lanceolate to linear-lanceolate, 10–15 cm long, margins entire. *Flowers* in axillary, umbel-like cymes; corolla reddish orange, 5-lobed, lobes 7–8 mm long, reflexed; corona yellowish orange, hoods 4–5 mm long, horns slightly longer; stamens 5. *Fruit* a follicle, spindle-shaped, to 10 cm long; seeds numerous, brown, each with a tuft of silky white hairs.

Comments: The eye-catching flowers of this species stand out in the fields and pastures of the highlands where it normally grows. The genus name is derived from *Asklepios*, the Greek god of medicine. This refers to the medicinal properties of certain members of the genus.

301. *Asclepias curassavica* (butterfly weed), flowers

Scientific Name: Bidens pilosa L. (Photo 302)
Common Names: Beggartick, Spanish needle

Family: Asteraceae (Sunflower)

Range: Native; also known from other tropical regions throughout the world, originally from tropical America.

Islands Inhabited: Fernandina, Floreana, Isabela (A,D,SN), San Cristóbal, Santa Cruz

Habitat: Moist uplands

Description: Annual herb to 1 m tall. *Leaves* opposite, variable, some simple, others pinnately compound, to ca. 10 cm long, margins serrate. *Flowers* in discoid or radiate heads; heads solitary or several, terminal or axillary. Discoid heads 5–7 mm across; flowers yellow, perfect. Radiate heads to 1 cm across; disc flowers yellow, perfect; ray flowers white, sterile. *Fruit* an achene, to 1 cm long, usually 2 awns on top with downward pointing barbs, 2–4 mm long; seed 1.

Comments: This herb is a common weed along roadsides and disturbed areas and is easily dispersed by means of its barbed fruits. In fact, it may well have arrived in the Galápagos attached to the feathers of a bird. *Bidens cynapiifolia* HBK., an introduced weed found only on San Cristóbal, differs in that each achene typically possesses 3–5 erect awns. *Bidens riparia* HBK., a native, also possesses 3–5 awns, but at least 1 is bent at a right angle to the body of the achene. It is known to inhabit Española, Fernandina, Floreana, Isabela (A,CA,D), Santa Cruz, and Santiago.

302. *Bidens pilosa* (beggartick), flowers

Scientific Name: Jaegeria gracilis Hook. f. (Photo 303)
Common Name: Galápagos jaegeria

Family: Asteraceae (Sunflower)

Range: Endemic

Islands Inhabited: Fernandina, Floreana, Isabela (A,CA,D,SN), San Cristóbal, Santa Cruz, Santiago

Habitat: Moist uplands

Description: Delicate, low-growing annual, stems often rooting at the nodes, stem tips curving upward, typically to 30 cm tall, occasionally taller. *Leaves* opposite, simple, lanceolate to ovate, 1–4 cm long, margins usually entire. *Flowers* in discoid or radiate heads; heads typically solitary, terminal or axillary. Discoid heads 4–6 mm across, flowers yellow, perfect. Radiate heads small, disc 4–6 mm across; disc flowers yellow, perfect; ray flowers yellow, pistillate. *Fruit* an achene, black, ca. 1 mm long; seed 1.

Comments: This is the only species of *Jaegeria* currently recognized in the Galápagos. Specimens of *J. crassa* Torres, mentioned in Wiggins and Porter (1971), are now simply considered to be more robust, forest-dwelling members of *J. gracilis* (van der Werff 1977).

303. *Jaegeria gracilis* (Galápagos jaegeria), flowers

Scientific Name: Pectis subsquarrosa (Hook. f.) Schultz-Bip. (Photo 304)
Common Name: Pectis

Family: Asteraceae (Sunflower)

Range: Endemic

Islands Inhabited: Baltra, Española, Fernandina, Floreana, Isabela (A,D), Pinzón, Rábida, San Cristóbal, Santa Cruz, Santa Fe, Santiago, Seymour, Islet(s)

Habitat: Arid lowlands

Description: Annual herb to 50 cm tall, much-branched, usually forming a somewhat rounded clump. *Leaves* opposite, simple, linear-oblong, typically 0.4–2.5 cm long, sometimes longer, lower surface with distinct oil glands, margins entire except near base, where there may be a few pairs of stiff bristles. *Flowers* in radiate heads ca. 6–7 mm across; heads borne near the stem tips; disc flowers yellow, perfect; ray flowers yellow above, reddish brown beneath, pistillate. *Fruit* an achene, black, 2.5–3.5 mm long, usually with 2 to many scales or bristles on top, 2–3 mm long, sometimes no scales or bristles; seed 1.

Comments: The genus name comes from the Latin *pecten*, meaning "comb." This refers to the stiff bristles on the leaf margins. All members of this genus produce a distinct aroma when their leaves are rubbed between the fingers.

304. *Pectis subsquarrosa* (pectis), flowers and fruits

Scientific Name: Pectis tenuifolia (DC.) Schultz-Bip. (Photos 305–6)
Common Name: Pectis

Family: Asteraceae (Sunflower)

Range: Endemic

Islands Inhabited: Fernandina, Floreana, Genovesa, Isabela (A,CA,D,SN), San Cristóbal, Santa Cruz, Santa Fe, Islet(s)

Habitat: Arid lowlands

Description: Perennial herb to 15 cm tall, much-branched, usually forming a somewhat rounded clump. *Leaves* opposite, simple, linear-filiform, typically 1.5–4 cm long, lower surface with distinct oil glands, margins entire except the lower halves, which are covered with stiff bristles. *Flowers* in radiate heads ca. 6–7 mm across; heads borne near the stem tips; disc flowers yellow, perfect; ray flowers yellow above, reddish brown beneath, pistillate. *Fruit* an achene, black, 3–4 mm long, usually with 2 to many scales or bristles on top, 3–5 mm long, sometimes no scales or bristles; seed 1.

Comments: This plant typically roots itself in crevices on barren lava. *Pectis linifolia* L., a native, differs from *P. subsquarrosa* and *P. tenuifolia* primarily in that its ray flowers are less conspicuous. In addition, its achenes typically have 2–4 obvious, somewhat spreading, often curved awns on top. It inhabits Baltra, Isabela (A,D), Pinta, Rábida, Santa Cruz, Santiago, and one or more of the smaller islands.

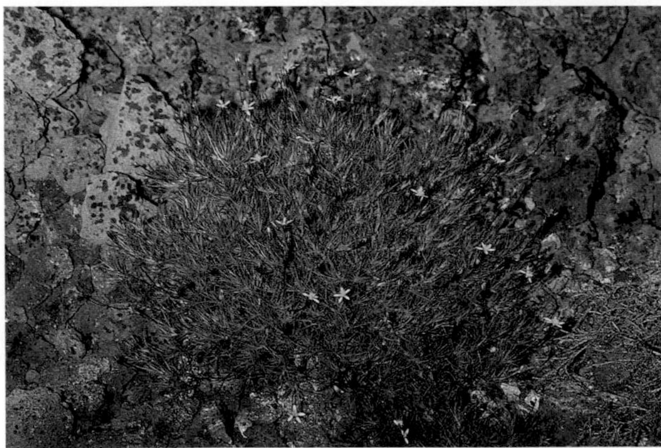

305. (left) *Pectis tenuifolia* (pectis), flowers

306. (above) *Pectis tenuifolia* (pectis), flowers

Scientific Name: Hypericum uliginosum Kunth (Photos 307–8)
Common Name: St. John's wort

Family: Clusiaceae (Garcinia)

Range: Native; also known from other regions of tropical America (Mexico to northern South America).

Islands Inhabited: Isabela (SN), San Cristóbal, Santa Cruz, Santiago

Habitat: Moist uplands

Description: Perennial herb to 50 cm tall, becoming somewhat woody at the base. *Leaves* opposite, simple, linear-lanceolate to ovate-lanceolate, 8–25 mm long, becoming reddish with age, with tiny, yellow, dot-shaped glands that turn black, margins entire. *Flowers* in terminal cymes 5–10 cm long; corolla yellow with reddish tinge, petals 5, 3–3.5 mm long; stamens 9–10. *Fruit* a capsule, narrowly ovoid, ca. 2.5–3 mm long; seeds numerous, reddish brown.

Comments: All Galápagos members of this species represent var. *pratense* (Cham. & Schlecht.) Keller. The genus name comes from the Greek *hyper*, "above," and *eikon*, "picture." Apparently, in ancient times St. John's wort was placed above pictures to ward off evil spirits.

307. (left) *Hypericum uliginosum* var. *pratense* (St. John's wort), flowers

308. (above) *Hypericum uliginosum* var. *pratense* (St. John's wort), flowers

Scientific Name: Calceolaria meistantha Pennell (Photo 309)
Common Name: Slipper flower

Family: Scrophulariaceae (Figwort)

Range: Native; also known from Colombia, mainland Ecuador, and Peru.

Islands Inhabited: Fernandina, Isabela (CA,SN,W)

Habitat: Moist uplands

Description: Annual herb typically to 15 cm tall, occasionally taller, stems covered with glandular hairs. *Leaves* opposite, simple; blade ovate, typically 7–20 mm long, occasionally longer, both surfaces somewhat hairy, margins crenate to minutely serrate. *Flowers* axillary, solitary; corolla yellow with 2 brownish spots inside, 5–6 mm long, 2-lipped, lower lip slipper-shaped; stamens 2. *Fruit* a capsule, ovoid, 5–6 mm long; seeds numerous.

Comments: Although rarely seen, the beautiful flowers of this species are a treat to behold. *Calceolaria mexicana* Benth. differs primarily in having leaves that vary from ovate-shaped with dentate margins to deeply pinnately lobed, with dentate or serrate lobe margins (Hamann 1979). It is introduced and known only from Santa Cruz.

309. *Calceolaria meistantha* (slipper flower), flowers

Scientific Name: Tribulus cistoides L. (Photo 310)
Common Names: Puncture weed, goat's head, *cacho de chivo*

Family: Zygophyllaceae (Caltrop)

Range: Native; also known from other tropical regions throughout the world, originally from Africa.

Islands Inhabited: Baltra, Daphne, Darwin, Española, Fernandina, Floreana, Isabela (D,SN), Pinta, Rábida, San Cristóbal, Santa Cruz, Santiago, Seymour, Islet(s)

Habitat: Arid lowlands

Description: Low-growing perennial herb, somewhat woody at the base, stems to 75 cm long, covered with fine hairs. *Leaves* opposite, those of each pair unequal in size, even-pinnately compound, 2.5–8.5 cm long; leaflets usually 6–8 pairs, occasionally to 10, oblong to elliptic, largest ones 6–21 mm long, covered with tiny white hairs. *Flowers* appearing axillary, solitary, 1.5–2.5 cm across; corolla yellow, petals 5, 7–17 mm long; stamens 10. *Fruit* a schizocarp, grayish, hairy, to ca. 1.5 cm wide, with 5 sections, each section usually with 2 large spines (5–7 mm long) and 2 smaller spines; seeds numerous.

Comments: The genus name is a Latinized form of the Greek *tribolos*, "caltrop." A caltrop is an iron ball with four spikes arranged in such a way that when three are touching the ground, the other is pointing upward. Such devices were formerly used in warfare to slow the advance of troops. Thus, the genus name refers to the individual fruit sections, each of which has 4 spines. These occasionally stick to animals, and they may well have arrived in the archipelago attached to the feathers of some unsuspecting bird. Additionally, they are known to hitch rides on hikers' boots and the tires of cars, trucks, and airplanes. This explains the frequency of puncture weed near trails, roadsides, and airports. The seeds of this plant, although housed inside an extremely hard fruit, are known to be a source of food for medium ground finches (*Geospiza fortis*) and large ground finches (*G. magnirostris*) on Daphne (Grant 1986).

Tribulus terrestris L. (puncture weed), also a native, differs primarily in having smaller flowers (5–10 mm across). It occurs on Española, Floreana, Genovesa, Isabela (D,SN), Santa Cruz, Santiago, and one or more of the smaller islands. It too is commonly observed in disturbed areas such as roadsides and on trails.

A similar genus in the same family, *Kallstroemia*, may be differentiated from *Tribulus* by its fruits. These are divided into 10 sections rather than 5 and possess bumps instead of spines. *Kallstroemia adscendens* (Andersson) B. L. Rob., an endemic, is found on Darwin, Española, Fernandina, Floreana, Isabela (A,D,E), Pinzón, San Cristóbal, Santa Fe, and one or more of the smaller islands.

310. *Tribulus cistoides* (puncture weed), flowers and fruit

See also

Euphorbia cyathophora (p. 233)
Mirabilis jalapa (p. 299)
Portulaca howellii (p. 211)
P. oleracea (p. 212)

Herbs with Opposite Leaves and Pink, Red, or Purple Flowers

Scientific Name: Blechum pyramidatum (Lam.) Urb. (Photo 311)
Common Name: Blechum

Family: Acanthaceae (Acanthus)

Range: Native; also known from other tropical regions throughout the world, originally from Mexico to northern South America.

Islands Inhabited: Floreana, Isabela (CA,SN), Pinta, San Cristóbal, Santa Cruz, Santiago

Habitat: Arid lowlands and moist uplands

Description: Low-growing perennial herb to 70 cm long, stem tips curving upward. *Leaves* opposite, simple; blade ovate to lanceolate, 2–7 cm long, often covered with fine hairs, margins mostly entire. *Flowers* in dense, 4-sided, upright spikes 3–6 cm long, each flower with a hairy bract beneath; corolla purplish or white, tubular with 5 lobes, 12–15 mm long; stamens 4. *Fruit* a capsule, oblong, to 6 mm long; seeds numerous, brown.

Comments: This species produces interesting seeds that appear smooth when dry but, upon wetting, are seen to possess mucilagenous hairs on the edges. Wiggins and Porter (1971) list this plant as *B. brownei* Juss. forma *puberulum* Leonard.

311. *Blechum pyramidatum* (blechum), flowers

Scientific Name: Justicia galapagana Lindau (Photo 312)
Common Name: Galápagos justicia

Family: Acanthaceae (Acanthus)

Range: Endemic

Islands Inhabited: Fernandina, Isabela (A,CA,SN,W), Pinta, Santa Cruz, Santiago

Habitat: Moist uplands

Description: Herb to 1 m tall, stems somewhat hairy, occasionally rooting at the nodes. *Leaves* opposite, simple; blade lanceolate to ovate, to 5 cm long, somewhat hairy, margins entire. *Flowers* in axillary panicles covered with sticky hairs; corolla typically purple with white markings inside the throat, occasionally completely white, tubular with 2 distinct lips, to 14 mm long, upper lip 2-lobed, lower lip 3-lobed; stamens 2. *Fruit* a capsule, oblong, to 12 mm long; seeds 4, brownish, covered with fine hairs.

Comments: This herb grows well both in the shade and in open, sunny areas. Its flowers are a favorite source of nectar for syrphid flies (*Toxomerus crockeri*). In fact, these tiny insects almost disappear from view as they force their way deep into the flower's throat to feed.

312. *Justicia galapagana* (Galápagos justicia), flower

Scientific Name: Sesuvium portulacastrum (L.) L.

(Photos 313–14; see also Photo 18)

Common Names: Common carpetweed, sea purslane

Family: Aizoaceae (Carpetweed)

Range: Native; also known from other tropical regions throughout the world.

Islands Inhabited: Fernandina, Isabela (A,CA), Marchena, Santa Cruz, Santa Fe, Santiago, Seymour

Habitat: Coastal zone

Description: Low-growing perennial herb, sometimes with a woody base, stems fleshy. *Leaves* opposite, simple, oblanceolate or oblong, 3–5 cm long, fleshy. *Flowers* axillary, solitary; calyx pinkish, 5-lobed, lobes 6–7 mm long; corolla absent; stamens numerous. *Fruit* a capsule, ovoid with caplike lid; seeds numerous, shiny black, with smooth seed coats.

Comments: Common carpetweed generally forms a dense cover over rocky and sandy terrain. During the cool season it often turns an unmistakable orange-red color. Its flowers are frequently visited by the Galápagos carpenter bee (*Xylocopa darwini*) and the Galápagos sulfur butterfly (*Phoebis sennae*).

313. *Sesuvium portulacastrum* (common carpetweed)

314. *Sesuvium portulacastrum* (common carpetweed), flower

Scientific Name: Catharanthus roseus (L.) G. Don (Photo 315)
Common Name: Madagascar periwinkle

Family: Apocynaceae (Dogbane)

Range: Cultivated; also known from other tropical regions throughout the world, originally from Madagascar.

Islands Inhabited: Isabela (SN), Santa Cruz

Habitat: Arid lowlands

Description: Perennial herb to 60 cm tall, sometimes woody at the base, stems containing a milky sap. *Leaves* opposite, simple, oblong-elliptic to obovate, 3–5 cm long, margins entire. *Flowers* axillary, in pairs; corolla pink with a reddish purple center and white throat, or completely white, salverform with 5 lobes, to 4 cm across; stamens 5. *Fruit* a follicle, to 4 cm long; seeds numerous.

Comments: This plant is commonly found around homes in the archipelago and is often used for centerpieces on restaurant tables. It is thought to have medicinal properties effective against some forms of cancer; however, it is poisonous if eaten.

315. *Catharanthus roseus* (Madagascar periwinkle), flowers

Scientific Name: Ageratum conyzoides L. (Photo 316)
Common Name: Ageratum

Family: Asteraceae (Sunflower)

Range: Native; also known from other tropical regions throughout the world, originally from Central and South America.

Islands Inhabited: Floreana, Isabela (A,SN), Pinta, San Cristóbal, Santa Cruz, Santiago

Habitat: Moist uplands

Description: Annual herb to 1 m tall, branching, stems somewhat hairy. *Leaves* opposite, simple; blade ovate, typically to ca. 7 cm long, glandular dots often present, veins or entire blade sometimes purple, margins coarsely serrate to crenate. *Flowers* in discoid heads, disc 2–5 mm across; heads numerous in tight clusters, borne near the branch tips; flowers typically blue to purplish or pink, occasionally white, perfect. *Fruit* an achene, black, 1.5 mm long, with 5 awn-tipped scales on top, 1.3 mm long; seed 1.

Comments: One of the easily identifiable characteristics of this plant is its unpleasant odor. The genus name comes from the Greek *a*, "not," and *geras*, "age," alluding to the length of time these plants keep their flowers.

316. *Ageratum conyzoides* (ageratum), flowers

Scientific Name: Kalanchoe pinnata (Lam.) Pers. (Photo 317)
Common Names: Air plant, *hoja del aire*

Family: Crassulaceae (Orpine)

Range: Cultivated escape; also known from other tropical regions throughout the world.

Islands Inhabited: Floreana, Isabela (SN), Santa Cruz, San Cristóbal

Habitat: Arid lowlands and moist uplands

Description: Perennial to 2 m tall, rarely branching, stems fleshy, becoming slightly woody at the base. *Leaves* opposite, the lower and upper ones simple, the middle ones usually odd-pinnately compound; leaflets 3–5, elliptic or obtuse, 5–20 cm long, somewhat fleshy, margins crenate. *Flowers* in terminal cymes to ca. 80 cm long; calyx reddish purple and pale greenish yellow, cylindrical with 4 lobes, 2.5–4.5 cm long; corolla reddish brown, tubular with 4 lobes, 3–6 cm long; stamens 8. *Fruit* a follicle; seeds numerous.

Comments: Air plant is normally cultivated for landscaping, but in parts of the archipelago it has escaped to open, disturbed areas such as roadsides and fields. One of this plant's well-known features is its ability to reproduce vegetatively by forming tiny rootlets in the notches of its leaf margins. When the leaves fall to the ground, they can immediately begin to form new adults. This is such a successful strategy that they can eventually dominate the areas they inhabit. This is happening in parts of the highlands of Volcán Sierra Negra (Isabela). This species can also be an extreme nuisance to greenhouse owners, who find that vigilance and frequent weeding are necessary to control its spread.

317. *Kalanchoe pinnata* (air plant), flowers

Scientific Name: Teucrium vesicarium Mill. (Photo 318)
Common Name: Germander

Family: Lamiaceae (Mint)

Range: Native; also known from other regions of tropical America (Mexico to Argentina).

Islands Inhabited: Floreana, Isabela (A,CA,D,SN), San Cristóbal, Santa Cruz

Habitat: Moist uplands

Description: Perennial herb to 1.2 m tall, stems square in cross section. *Leaves* opposite, simple; blade ovate, 4–12 cm long, both surfaces somewhat hairy, margins crenate-serrate. *Flowers* in terminal spikelike racemes to 15 cm long, each flower with a bract beneath; corolla purplish to white, tubular with 2 lips, 9–13 mm long, upper lip 2-lobed, lower lip 3 lobed, the middle lobe of the lower lip much larger than the lateral lobes, outer surface of entire corolla and inner surface of lower lip covered with tiny glandular hairs; stamens 4. *Fruit* consisting of 4 nutlets, reddish brown, 2.5–3 mm long, enclosed in an inflated calyx covered with short, curled hairs, 4–5 mm in diameter; seeds 1 per nutlet.

Comments: This species is sometimes mistaken for *Priva lappulacea* (priva) (see Photo 300), which also has squarish stems, similar leaves, and an inflated calyx around each of its fruits. However, the flowers of *P. lappulacea* are much smaller (ca. 3.5 mm long). If no flowers are available, then look at the hairs on an inflated calyx: Those of *P. lappulacea* have hooked tips.

318. *Teucrium vesicarium* (germander), flowers

Scientific Name: Boerhaavia caribaea Jacq. (Photo 319)
Common Name: Boerhaavia

Family: Nyctaginaceae (Four-o'clock)

Range: Native; also known from other regions of tropical America (southern United States to Bolivia).

Islands Inhabited: Daphne, Española, Fernandina, Floreana, Genovesa, Isabela (D,SN), Pinta, San Cristóbal, Santa Cruz, Santa Fe, Santiago, Islet(s)

Habitat: Arid lowlands

Description: Prostrate to erect perennial herb, stems covered with tiny glandular hairs. *Leaves* opposite, simple, broadly ovate to roundish, 2–4 cm long, both surfaces with glandular hairs, margins slightly wavy, with fine hairs. *Flowers* in axillary clusters, the branches of the inflorescences covered with fine glandular hairs; calyx reddish purple, campanulate with 5 lobes, ca. 1.5 mm long; corolla absent; stamens 1–3. *Fruit* an achene, obovoid with 5 longitudinal grooves, ca. 2.5 mm long, covered with fine, sticky, glandular hairs; seed 1.

Comments: The genus name honors Hermann Boerhaave (1668–1738), a botanist at the University of Leiden, Netherlands. Often its fruits are observed clinging to one's clothing by the sticky, glandular hairs.

Boerhaavia coccinea Mill. differs in having smooth inflorescence branches. It is an introduced species known only from Floreana. *Boerhaavia erecta* L. differs in having both smooth inflorescence branches and smooth fruits. In addition, it has tiny red dots on its leaves, and flowers that are pink or white. This native occurs on Daphne, Española, Fernandina, Floreana, Isabela (A,D,SN), Rábida, Santa Cruz, Santa Fe, Santiago, and one or more of the smaller islands.

319. *Boerhaavia caribaea* (boerhaavia), flowers and fruits

Scientific Name: Commicarpus tuberosus (Lam.) Standl. (Photos 320–21)
Common Name: Wartclub

Family: Nyctaginaceae (Four-o'clock)

Range: Native; also known from mainland Ecuador and Peru.

Islands Inhabited: Floreana, Isabela (A,CA,SN), Pinzón, Rábida, San Cristóbal, Santa Cruz, Santiago, Islet(s)

Habitat: Arid lowlands

Description: Prostrate to erect herb, much-branched. *Leaves* opposite, simple, broadly ovate to somewhat cordate or roundish, 3–6 cm long, margins entire. *Flowers* in terminal umbels; calyx purplish or pink, funnelform with 5 lobes, 8–10 mm long; corolla absent; stamens 2–3. *Fruit* an achene, club-shaped, 6–10 mm long, with 10 faint longitudinal grooves or lines, several wartlike glands near the top; seed 1.

Comments: The wartlike glands of these fruits are extremely sticky and will adhere to any animals that pass by. They also attach themselves to the socks and pants of many tourists.

320. (above) *Commicarpus tuberosus* (wartclub), flowers

321. (left) *Commicarpus tuberosus* (wartclub), flowers and fruits

Scientific Name: Verbena litoralis HBK. (Photo 322)
Common Name: Vervain

Family: Verbenaceae (Vervain)

Range: Native; also known from other tropical regions throughout the world.

Islands Inhabited: Floreana, Isabela (SN), San Cristóbal, Santa Cruz

Habitat: Moist uplands

Description: Perennial herb to 2 m tall, stems 4-angled, sometimes becoming somewhat woody at the base. *Leaves* opposite, simple, lanceolate to oblanceolate, 3–10 cm long, both surfaces somewhat hairy, upper surface sometimes rough, margins coarsely serrate on upper two-thirds. *Flowers* in terminal spikes to 8 cm long, each flower with a bracteole beneath; corolla purplish, bluish, or pink; salverform with 5 lobes, ca. 2.5–3 mm across, somewhat 2-lipped; stamens 4. *Fruit* a schizocarp, oblong, ca. 2 mm long, mostly enclosed in calyx, separating into 4 sections when mature; seed 1 per section, each triangular in cross section.

Comments: This species is also known from Hawaii, and Wagner et al. (1990) state that it has been used there as a treatment for cuts, bruises, and fractures. *Verbena grisea* B. L. Rob. & Greenm. (lobed vervain), an endemic found only on Pinzón, is considered vulnerable. It differs from all other members of the genus in having leaves that are bipinnately lobed.

Verbena townsendii Svens. (Townsend's vervain) is an endemic found on Fernandina, Isabela (SN), Pinta, and Santa Cruz. According to van der Werff (1977), it includes those plants listed by Wiggins and Porter (1971) as *V. galapagosensis* Moldenke, *V. glabrata* var. *tenuispicata* Moldenke, and *V. stewartii* Moldenke.

Verbena sedula Moldenke, also an endemic species, may be divided into three varieties. These are var. *darwinii* Moldenke (Santiago), var. *fournieri* Moldenke (San Cristóbal), and var. *sedula* (Santa Cruz).

The remaining member of this genus, *V. brasiliensis* Vell. (Brazilian vervain), is an introduced species. It is restricted to San Cristóbal and has been collected along the road near Tres Palos (van der Werff 1977).

322. *Verbena litoralis* (vervain), flowers

See also

Chamaesyce hirta (p. 251)
Cuphea carthagenensis (p. 253)
Mirabilis jalapa (p. 299)
Porophyllum ruderale (p. 217)
Priva lappulacea (p. 257)
Scoparia dulcis (p. 255)
Trianthema portulacastrum (p. 246)

Herbs with Opposite Leaves and Blue Flowers

See

Ageratum conyzoides (p. 271)
Priva lappulacea (p. 257)
Verbena litoralis (p. 276)

Herbs with Opposite Leaves and Green Flowers

Scientific Name: Pilea baurii B. L. Rob. (Photo 323)
Common Names: Galápagos pilea, *ortiga*

Family: Urticaceae (Nettle)

Range: Endemic

Islands Inhabited: Española, Fernandina, Floreana, Isabela (A,SN), Pinta, Pinzón, San Cristóbal, Santa Cruz, Santiago

Habitat: Arid lowlands and moist uplands

Description: Annual or short-lived perennial herb to 60 cm tall, stems somewhat fleshy, often reddish. *Leaves* opposite, simple; petiole often red, blade ovate, 2–10 cm long, both surfaces often containing tiny cystoliths (calcium carbonate deposits), margins serrate. *Flowers* unisexual (plants monoecious) in axillary panicles. Staminate flowers near the bottom of each panicle; calyx green, 4-lobed, lobes 1–1.2 mm long; corolla absent; stamens 4. Pistillate flowers near the top of each panicle; calyx green, 2-lobed, lobes 0.4–1.4 mm long; corolla absent. *Fruit* an achene, elliptic-obovate, 0.3–0.4 mm long; seed 1.

Comments: This species typically grows in relatively moist, shaded areas. According to Wiggins and Porter (1971), tortoises near El Chato (Santa Cruz) are known to feed on this herb.

Pilea microphylla (L.) Liebm. (artillery plant) and *P. peploides* (Gaud.) Hook. & Arn. differ from *P. baurii* in having leaf margins that are not serrate. In addition, their leaf blades are much smaller, usually less than 1.5 cm long. They differ from each other in that *P. microphylla* has obovate to oblanceolate leaves that are 2–15 mm long, while *P. peploides* has ovate to roundish leaves that are 1–6 mm long. *Pilea microphylla* is an introduced species known only from Santa Cruz. Its common name comes from the fact that its pollen is forcefully discharged, or shot, from the mature anthers. *Pilea peploides* is a native that occurs on Fernandina, Floreana, and Isabela (A,D,SN).

323. *Pilea baurii* (Galápagos pilea), flowers

See also

Chamaesyce hirta (p. 251)
Porophyllum ruderale (p. 217)

Herbs with Whorled Leaves and White Flowers

Scientific Name: Mollugo flavescens Andersson (Photos 324–25)
Common Name: Mollugo

Family: Molluginaceae (Mollugo)

Range: Endemic

Islands Inhabited: Baltra, Daphne, Española, Fernandina, Floreana, Genovesa, Isabela (A,CA,D,SN,W), Marchena, Pinta, Pinzón, Rábida, San Cristóbal, Santa Cruz, Santa Fe, Santiago, Islet(s)

Habitat: Arid lowlands

Description: Low-growing to erect annual herb to 40 cm tall. *Leaves* typically in whorls, simple, linear, oblanceolate, spatulate, or obovate, 5–20 mm long, smooth, margins entire. *Flowers* axillary, typically 3–8 per cluster; sepals 4–5, 1.5–2 mm long, whitish with green veins; corolla absent; stamens typically 3–6, occasionally 8. *Fruit* a capsule; seeds numerous, swollen-reniform, dark reddish brown or black, shiny, covered with tiny bumps.

Comments: This species includes four subspecies. *Mollugo flavescens* subsp. *striata* (J. T. Howell) Eliasson is known only from Genovesa and Wolf and does not share these islands with any other members of the genus. The other three subspecies, however, do overlap somewhat in their distribution and are extremely difficult to distinguish without careful study. *Mollugo flavescens* subsp. *insularis* (J. T. Howell) Eliasson is considered rare and is known only from Floreana and San Cristóbal. *Mollugo flavescens* subsp. *flavescens* occurs on Daphne, Española, Rábida, San Cristóbal, and Santa Cruz. *Mollugo flavescens* subsp. *gracillima* (Andersson) Eliasson (Photos 324–25) is found on Baltra, Bartolomé, Española, Fernandina, Floreana, Isabela (A,CA,D,SN,W), Marchena, Pinta, Pinzón, Rábida, San Cristóbal, Santa Cruz, Santa Fe, Santiago, and one or more of the smaller islands. *Mollugo flavescens* subsp. *insularis* differs from the other two subspecies in that its seeds are obviously ridged on one side. The seeds of *M. flavescens* subsp. *flavescens* and subsp. *gracillima*, although similar in not having ridges, do differ slightly. Each seed of the former typically retains part of the stalk that held it inside the capsule, while seeds of the latter do not.

Mollugo floriana (B. L. Rob.) J. T. Howell, also an endemic, differs from *M. flavescens* in that its leaves are usually 12–32 mm long and its flowers typically have 7–8 stamens. It includes three subspecies. *Mollugo floriana* subsp. *floriana* is considered rare but may be found on Floreana, Santa Fe, and one or more of the smaller islands. *Mollugo floriana* subsp. *gypsophiloides* J. T. Howell is also rare and is known only from Pinzón. *Mollugo floriana* subsp. *santacruziana* (Christoph.) Eliasson occurs on San Cristóbal, Santa Cruz, and Santiago.

Mollugo snodgrassii B. L. Rob. (Snodgrass' mollugo) (Photo 326) is an endemic that inhabits the arid lowlands of Fernandina and Isabela (A,CA,D,E,SN,W). It is the only member of the genus with sepals that are 3 mm or more in length, and flower stalks that are regularly 1 cm or more in length. This plant is common on relatively uncolonized lava fields such as those at Volcán Chico (Isabela).

Mollugo cerviana (L.) Ser. is a native found on Baltra, Isabela (A,D), Pinta, Santiago, and one or more of the smaller islands. It also occurs in other tropical regions throughout the world. It stands out from all other species in the genus by producing seeds that are semicircular in shape, rather than reniform.

Mollugo crockeri J. T. Howell (Crocker's mollugo) is an endemic known only from Santiago, and its status is considered vulnerable. This is the only member of the genus that possesses obvious glandular hairs on its stems, leaves, and inflorescences.

324. (above) *Mollugo flavescens* subsp. *gracillima* (mollugo)

325. (left) *Mollugo flavescens* subsp. *gracillima* (mollugo), flowers

326. *Mollugo snodgrassii* (Snodgrass' mollugo), flowers

Herbs with Whorled Leaves and Green Flowers

Scientific Name: Peperomia galapagensis Miq. (Photo 327)
Common Name: Galápagos peperomia

Family: Piperaceae (Pepper)

Range: Endemic

Islands Inhabited: Floreana, Isabela (A,CA,SN), Pinta, Pinzón, San Cristóbal, Santa Cruz, Santiago

Habitat: Moist uplands

Description: Herbaceous perennial to 15 cm tall, growing in clumps on ground, rocks, and trees, stems densely covered with fine hairs, stem tips bending upward. *Leaves* typically in whorls of 3–4 (occasionally more), simple; blade oval-oblong to almost spatulate, 5–8 mm long, 1-nerved, somewhat fleshy to leathery. *Flowers* in terminal and axillary spikes to 15 mm long, each flower with a roundish bract beneath, greenish; calyx absent; corolla absent; stamens 2. *Fruit* drupelike, roundish, 0.5–0.7 mm long; seed 1.

Comments: This species includes two varieties. *Peperomia galapagensis* var. *galapagensis* (Photo 327) has leaves that are smooth or only slightly hairy. It occurs on Floreana, Isabela (SN), Pinta, Pinzón, San Cristóbal, Santa Cruz, and Santiago. *Peperomia galapagensis* var. *ramulosa* (Andersson) Yuncker has leaves that are densely covered with tiny hairs. It is known from Floreana, Isabela (A,CA,SN), Pinta, and Santa Cruz.

Peperomia galioides HBK. is a native found on Fernandina, Isabela (A,SN), Pinta, Pinzón, San Cristóbal, Santa Cruz, and Santiago. It has densely hairy stems that grow to 1 m in length. Its leaves, which may reach a length of up to 30 mm, are arranged in whorls of 3–5 (occasionally more), and each is 3- to 5-nerved.

Peperomia obtusilimba C. DC. is a rare endemic found on Fernandina, Isabela (D), Pinta, Santa Cruz, and Santiago. It also has 3–5 leaves per node, each of which is 3- to 5-nerved. However, it differs from *P. galioides* in that the length of its leaf blades generally does not exceed 11 mm. In addition, its stems are less hairy and only occasionally reach a maximum length of 15 cm.

Peperomia petiolata Hook. f. and *P. tequendamana* Trel. typically have opposite leaves, although some alternate leaves are produced as well. The former is an endemic known from Fernandina, Floreana, Isabela (A,CA,SN,W), Pinta, Pinzón, San Cristóbal, Santa Cruz, and Santiago, while the latter is a native that occurs on Isabela (A), Santa Cruz, and Santiago. These two species differ in that the leaves of *P. tequendamana* possess numerous tiny black dots on their lower surface, while those of *P. petiolata* do not (van der Werff 1977). Both are considered rare.

327. *Peperomia galapagensis* var. *galapagensis* (Galápagos peperomia), flowers and fruits

Herbs with Clustered Leaves and White Flowers

See

Musa x *paradisiaca* (p. 187)
Tiquilia darwinii (p. 183)
T. galapagoa (p. 182)
T. nesiotica (p. 183)

Herbs with Clustered Leaves and Pink, Red, or Purple Flowers

See

Elytraria imbricata (p. 216)

Herbs with Clustered Leaves and Blue Flowers

See

Elytraria imbricata (p. 216)

Herbs with Basal Leaves and White Flowers

Scientific Name: Furcraea hexapetala (Jacq.) Urb. (Photos 328–29)
Common Names: Cuban hemp, *cabuya*

Family: Agavaceae (Agave)

Range: Cultivated escape; also known from other regions of tropical America (West Indies to northwestern South America).

Islands Inhabited: Floreana, Isabela (SN), San Cristóbal, Santa Cruz

Habitat: Arid lowlands and moist uplands

Description: Perennial herb with short underground stem. *Leaves* radiating upward and outward, simple; blade lanceolate, to ca. 1 m long, possessing a sharp tip 2–3 cm long, margins armed with curved prickles to 8 mm long. *Flowers* in large panicle at the top of a single stalk, panicle to 2 m tall, flowerless portion of stalk 5–7 m tall; tepals 6, white or greenish white, 2–2.5 cm long; stamens 6. *Fruit* a capsule, ovoid to cylindrical; seeds numerous, flat.

Comments: An interesting facet of Cuban hemp's life history is that it produces bulblets in addition to seeds. These miniature bulbs are produced on the same panicle as the flowers. When they fall to the ground, each can start a new individual. This is truly an impressive plant, but care must be taken when approaching it for closer inspection, as the leaves are well armed. According to Stewart (1911), settlers on San Cristóbal used its fibers to make rope. This species is listed as *F. cubensis* (Jacq.) Vent. in Wiggins and Porter (1971).

328. *Furcraea hexapetala* (Cuban hemp), flowers

329. *Furcraea hexapetala* (Cuban hemp), flowers

Scientific Name: Lithophila radicata (Hook. f.) Standl. (Photos 330–31)
Common Name: Lithophila

Family: Amaranthaceae (Amaranth)

Range: Endemic

Islands Inhabited: Española, Floreana, Pinzón, San Cristóbal, Islet(s)

Habitat: Coastal zone

Description: Perennial herb to 25 cm tall, numerous stems, often turning reddish. *Leaves* mainly basal, simple, quill-like, 5–15 cm long, 1.5–2 mm wide, stem leaves much reduced. *Flowers* in terminal spikes 8–20 mm long, each flower with 1 bract and 2 bracteoles beneath, bract and bracteoles white; calyx whitish, sepals 5, ca. 3 mm long; corolla absent; stamens 2, staminodia 3. *Fruit* a utricle; seed 1, reddish brown.

Comments: Lithophila subscaposa (Hook. f.) Standl., also an endemic, is found in the highlands of Floreana, Pinzón, and Santiago. It differs primarily in having basal leaves that are oblong-lanceolate, blunt at the tip, and usually smaller (typically 5–8 cm long, 5–10 mm wide). Both species are considered rare.

330. *Lithophila radicata* (lithophila), flowers

331. *Lithophila radicata* (lithophila), flowers

Scientific Name: Crinum latifolium L. (Photo 332)
Common Names: Crinum lily, spider lily, *lirio de cinta*

Family: Amaryllidaceae (Amaryllis)

Range: Cultivated; also known from other tropical regions throughout the world, originally from tropical Asia and Africa.

Islands Inhabited: Santa Cruz

Habitat: Arid lowlands

Description: Perennial herb to 1 m tall. *Leaves* basal, simple, straplike, to 1 m long, margins entire. *Flowers* in terminal umbels on leafless stalks; tepals 6, white inside, reddish purple and white outside, 8–10 cm long; stamens 6, filaments red. *Fruit* a capsule; seeds numerous.

Comments: This beautiful ornamental may be found growing next to homes and shops in Puerto Ayora. The genus derives its name from the Greek *krinon*, which means "lily." However, its flower differs from that of a true lily.

332. *Crinum latifolium* (crinum lily), flowers

Scientific Name: Tillandsia insularis Mez in DC. (Photo 333)
Common Names: Galápagos tillandsia, *huicundu*

Family: Bromeliaceae (Pineapple)

Range: Endemic

Islands Inhabited: Fernandina, Floreana, Isabela (A,CA,D,E,SN,W), Pinzón, San Cristóbal, Santa Cruz, Santiago

Habitat: Arid lowlands and moist uplands

Description: Short-stemmed perennial herb, usually living on other plants. *Leaves* forming a radiating cluster near the bottom, simple, strap-shaped, 13–45 cm long, often reddish near the base, margins entire. *Flowers* in short spikes at the top of a long stalk, short spikes 2–9 cm long, flowerless portion of stalk to 75 cm long, bracts present; corolla white, petals 3, ca. 5 mm long; stamens 6. *Fruit* a capsule, to 2.3 cm long; seeds numerous.

Comments: Galápagos tillandsia is an epiphyte, which means that it grows upon another plant. However, it does not function as a parasite. Instead, it removes moisture from the atmosphere through specialized cells in its leaves. According to Wiggins and Porter (1971), two varieties of this species exist: *T. insularis* var. *insularis* and var. *latilamina* Gilmartin. However, more recent studies by van der Werff (1977) suggest that the morphological variation observed among members of this species does not merit their subspecific recognition.

333. *Tillandsia insularis* (Galápagos tillandsia)

Herbs with Basal Leaves and Yellow or Orange Flowers

Scientific Name: Hypoxis decumbens L. (Photo 334)
Common Name: Yellow star-grass

Family: Hypoxidaceae (Star-grass)

Range: Native; also known from other regions of tropical America (Mexico and the West Indies to central South America).

Islands Inhabited: Floreana, Isabela (A,CA,SN), Pinta, San Cristóbal, Santa Cruz, Santiago

Habitat: Moist uplands

Description: Stemless herb to 30 cm tall. *Leaves* basal, simple, linear-lanceolate and curved, 5–30 cm long, margins hairy. *Flowers* produced on stalks to 20 cm long; tepals 6, yellow on the inside, green and hairy on the outside, ovate-lanceolate, 4–6 mm long; stamens 6. *Fruit* a capsule, oblong-ellipsoid, 7–12 mm long; seeds numerous, black, appearing somewhat warty.

Comments: Sisyrinchium macrocephalum R. C. Graham (Iridaceae), a rare endemic, shows a passing resemblance to *Hypoxis decumbens.* However, its flowers have only 3 stamens. It is known from Floreana, Isabela (CA,SN), San Cristóbal, and Santa Cruz.

334. *Hypoxis decumbens* (yellow star-grass), flower

Herbs with Basal Leaves and Pink, Red, or Purple Flowers

Scientific Name: Ananas comosus (L.) Merr. (Photo 335)
Common Names: Pineapple, *piña*

Family: Bromeliaceae (Pineapple)

Range: Cultivated; also known from other tropical regions throughout the world, originally from tropical America.

Islands Inhabited: San Cristóbal, Santa Cruz

Habitat: Moist uplands

Description: Short-stemmed perennial herb. *Leaves* basal, simple, strap-shaped, somewhat limp, to ca. 1 m long, often reddish near the base, margins spiny to serrate. *Flowers* in a spike, each flower with a bract beneath; corolla reddish purple, petals 3; stamens 6. *Fruit* a berry, all those of each spike fusing together into a single large multiple fruit, to ca. 30 cm long; seeds numerous, but so small as to be unnoticeable.

Comments: Pineapple is cultivated for its delicious fruit. The crown of leafy bracts on top of each fruit is caused by a continuation in the growth of the plant axis.

335. *Ananas comosus* (pineapple), fruit

See also

Crinum latifolium (p. 289)

Herbs with Basal Leaves and Green Flowers

Scientific Name: Plantago major L. (Photo 336)
Common Names: Common plantain, *llantén*

Family: Plantaginaceae (Plantain)

Range: Introduced; also known from other tropical regions throughout the world.

Islands Inhabited: Floreana, Isabela (A,SN), San Cristóbal, Santa Cruz, Santiago

Habitat: Arid lowlands and moist uplands

Description: Perennial herb. *Leaves* alternate but basally arranged, simple; blade broadly ovate, 5–25 cm long, margins wavy or toothed. *Flowers* in stalked spikes, stalks 6–40 cm tall, spikes 5–35 cm long, each flower with an elliptic bract beneath; corolla greenish, salverform with 4 lobes, lobes 1–1.5 mm long; stamens 4. *Fruit* a capsule, brownish, broadly ovoid to roundish, to 4 mm long, opening by means of a cap; seeds 6–18, reddish brown, sticky when wet.

Comments: Common plantain typically grows in disturbed sites such as roadsides and fields. *Plantago galapagensis* Rahn (Galápagos plantain), referred to as *P. paralias* Decne. var. *pumila* (Hook. f.) Wiggins in Wiggins and Porter (1971), is an endemic. It differs from *P. major* in that it is an annual and has elliptic to lanceolate leaves 2–10 cm long. In addition, it produces only 2 seeds in each capsule. It is considered rare but occurs on Floreana, Isabela (CA), San Cristóbal, Santa Cruz, and Santiago.

336. *Plantago major* (common plantain), flowers and fruits

Herbs with Basal Leaves and Brown Flowers

See

Cyperus anderssonii (p. 237)
C. ligularis (p. 238)

Vines with Alternate Leaves and White Flowers

Scientific Name: Ipomoea alba L. (Photo 337)
Common Name: Moon flower

Family: Convolvulaceae (Morning-glory)

Range: Native; also known from other tropical regions throughout the world, originally from tropical America, perhaps Mexico.

Islands Inhabited: Darwin, Fernandina, Isabela (A,CA,SN,W), Santiago

Habitat: Arid lowlands and moist uplands

Description: Perennial vine, stems to ca. 6 m long, containing a milky sap. *Leaves* alternate, simple; blade broadly ovate to ovate-lanceolate, 4–20 cm long, margins usually entire. *Flowers* axillary, solitary or in cymes; corolla white with a pale green "star" extending from the throat, cylindrical with a flat expanded top, 6–14 cm long, 9–15 cm across; stamens 5. *Fruit* a capsule, broadly ovoid, 3–3.5 cm long; seeds 4, dark brown to black.

Comments: This is one of 11 species of *Ipomoea* that occur in the Galápagos. The genus name is derived from the Greek *ips*, "worm," and *homoios*, "to resemble." This refers to the plant's twining appearance. In fact, it is often found growing on other plants. The specific epithet, *alba*, refers to the white color of the large fragrant flowers that open at night.

337. *Ipomoea alba* (moon flower), flowers

Scientific Name: Ipomoea habeliana Oliv. (Photo 338)
Common Names: Lava morning-glory, *soguilla*

Family: Convolvulaceae (Morning-glory)

Range: Endemic

Islands Inhabited: Española, Floreana, Genovesa, Isabela (W), Marchena, Pinta, Pinzón, Rábida, Santa Cruz, Islet(s)

Habitat: Arid lowlands

Description: Perennial vine, stems to 8 m long. *Leaves* alternate, simple; blade lanceolate to ovate-lanceolate, 6–15 cm long, usually standing upright on stems, margins entire. *Flowers* axillary, solitary or in cymes; corolla white, narrowly funnelform, 10–12 cm long, to ca. 6 cm across; stamens 5. *Fruit* a capsule, brown, broadly ovoid, 2–2.5 cm long; seeds to 4, brown, both lateral margins lined with light brown hairs.

Comments: This plant thrives on lava cliffs, in both shaded areas and full sunlight. Its flowers open in the late afternoon or early evening and last only until midmorning of the next day. By this time, the previously expanded top portion will have rolled up and withered. Hawk moths, attracted by the color and fragrance of these flowers, function as pollinators. Their reward is the copious nectar at the base of the corolla. Other insects, such as ants, are also attracted to these flowers. Even the Galápagos centipede (*Scolopendra galapagoensis*) has been observed deep inside a lava morning-glory flower.

338. *Ipomoea habeliana* (lava morning-glory), flowers

Scientific Name: Ipomoea linearifolia Hook. f. (Photo 339)
Common Name: Morning-glory

Family: Convolvulaceae (Morning-glory)

Range: Endemic

Islands Inhabited: Daphne, Fernandina, Genovesa, Isabela (A,E), Pinta, Rábida, Santa Cruz, Seymour, Wolf, Islet(s)

Habitat: Arid lowlands and moist uplands

Description: Perennial vine, older stems somewhat woody. *Leaves* alternate, simple; blade ovate to somewhat sagittate, 2–7 cm long. *Flowers* axillary, solitary or in cymes; corolla white or pink with a dark pink throat, broadly funnelform, 4.5–5 cm long, ca. 4.5–5 cm across; stamens 5. *Fruit* a capsule, ovoid, ca. 1.5 cm long; seeds 2–4.

Comments: The specific epithet of this plant, *linearifolia*, refers to the fact that the leaves are occasionally almost linear in shape. It often covers other vegetation or spreads extensively over lava fields.

339. *Ipomoea linearifolia* (morning-glory), flower

Scientific Name: Merremia aegyptica (L.) Urb. (Photo 340–41)
Common Name: Hairy merremia

Family: Convolvulaceae (Morning-glory)

Range: Native; also known from other tropical regions throughout the world, originally from tropical America.

Islands Inhabited: Daphne, Española, Genovesa, Isabela (A,CA,E,SN,W), Pinta, Pinzón, Rábida, Santa Cruz, Santa Fe, Santiago, Seymour, Islet(s)

Habitat: Arid lowlands

Description: Twining vine, stems to 4 m long, covered with hairs. *Leaves* alternate, palmately compound; leaflets 5, elliptic to obovate, 1.5–5 cm long, both surfaces covered with hairs. *Flowers* axillary, solitary or in cymes; calyx extremely hairy; corolla white, campanulate, 2–3 cm long, 2–2.5 cm across; stamens 5. *Fruit* a capsule, brown, somewhat spherical, 1–1.4 cm in diameter; seeds 4, brownish.

Comments: This vine is common in dry, disturbed sites, as well as in forested areas, and is often observed near towns and along roadsides. To see these flowers at their best, one must rise early, because the flowers open during the night or at daybreak but close by midmorning.

340. *Merremia aegyptica* (hairy merremia), covering a *Cordia lutea* (yellow cordia) tree during an El Niño year

341. *Merremia aegyptica* (hairy merremia), flowers and fruits

Scientific Name: Bougainvillea spectabilis Willd. (Photo 342)
Common Name: Bougainvillea

Family: Nyctaginaceae (Four-o'clock)

Range: Cultivated; also known from other regions of tropical America, originally from Brazil.

Islands Inhabited: Floreana, Santa Cruz

Habitat: Arid lowlands and moist uplands

Description: Woody perennial vine, stems with stout spines. *Leaves* alternate, simple, ovate to somewhat elliptic, to ca. 10 cm long, margins entire. *Flowers* in axillary clusters of threes, each flower with a purple, red, pink, or orange bract beneath, to ca. 3–4 cm long; calyx tubular with 5 lobes, to ca. 2 cm long, tube the same color as the bracts, lobes white, ca. 3 mm long; corolla absent; stamens 5–10. *Fruit* an achene, elongate, 5-ribbed; seed 1.

Comments: This species is commonly planted near porches for the shade it provides and for its lovely flowers, which may be found in additional colors to those mentioned above. Its fruits are dispersed by means of the winglike bracts to which they are attached. The genus was named for the French navigator, scientist, and explorer Louis Antoine de Bougainville (1729–1811).

Mirabilis jalapa L. (four-o'clock) (Photo 343) is another colorful member of this family. Its flowers are up to 5 cm long and come in shades of white, yellow, orange, pink, red, and purple. This species is grown as an ornamental on Isabela (SN), San Cristóbal, and Santa Cruz and has probably escaped cultivation. Its common name alludes to the fact that the flowers open in the late afternoon.

342. *Bougainvillea spectabilis* (bougainvillea), flowers

343. *Mirabilis jalapa* (four-o'clock), flowers and fruits

Scientific Name: Passiflora colinvauxii Wiggins (Photo 344)
Common Names: Colinvaux's passion flower, *granadilla silvestre*

Family: Passifloraceae (Passion flower)

Range: Endemic

Islands Inhabited: Santa Cruz

Habitat: Moist uplands

Description: Herbaceous climbing vine with axillary tendrils, stems slender. *Leaves* alternate, simple; blade 2-lobed, crescent-shaped, 7–16 cm wide. *Flowers* axillary, solitary or in pairs, 2–2.5 cm across; calyx pale green, sepals 5; corolla white, petals 5. In the middle of the flower is the androgynophore, a compound structure composed of 5 basal stamens and a 3-styled pistil. Between the petals and the androgynophore is the corona, which is composed of 2 series of purple and white filaments. Those of the outer series are longer than those of the inner series. *Fruit* a berry, green, elliptic-ovate, 2.5–4 cm long and 1–1.5 cm in diameter, seeds numerous, enclosed in a translucent mucilagenous substance.

Comments: This is one of seven species of *Passiflora* in the Galápagos. It occurs only in the highlands of Santa Cruz, where it normally climbs over other plants, including trees. The genus name is derived from the Latin *passio,* "passion," and *flos,* "flower." This is a reference to Christ's Passion, including the Crucifixion. In fact, each part of the flower has been used to symbolize an aspect of this event. For example, the combination of sepals and petals represent the 10 disciples present at the Crucifixion (Judas and Peter were absent); the 3 styles represent the nails in Christ's hands and feet; the 5 stamens represent the 5 wounds on Christ's body; and the corona represents the crown of thorns. Even the tendrils are said to symbolize the whips used by the Roman soldiers. The specific epithet of this rare species honors the ecologist Paul A. Colinvaux.

344. *Passiflora colinvauxii* (Colinvaux's passion flower), flower

Scientific Name: Passiflora edulis Sims. (Photos 345–46)
Common Names: Passion fruit, *maracuya*

Family: Passifloraceae (Passion flower)

Range: Cultivated escape; also known from other tropical regions throughout the world, originally from tropical America.

Islands Inhabited: Floreana, Santa Cruz

Habitat: Moist uplands

Description: Climbing vine with axillary tendrils, stems grooved and somewhat woody. *Leaves* alternate, simple; blade usually deeply 3-lobed but occasionally unlobed, ovate, 6–18 cm long, 7–20 cm wide, margins toothed. *Flowers* axillary, solitary or in pairs, to 9 cm across; calyx white, sepals 5; corolla white, petals 5; corona of numerous purple and white filaments. *Fruit* a berry, yellow, roundish, 5–9 cm in diameter; seeds numerous, enclosed in an orange to yellow mucilagenous substance.

Comments: This plant is widely cultivated for its fruit, which is used to make passion fruit juice. Visitors should try this drink when visiting one of the local restaurants. The plant itself may be seen in the highlands.

345. *Passiflora edulis* (passion fruit), flower

346. *Passiflora edulis* (passion fruit), fruits

Scientific Name: Passiflora foetida L. (Photo 347)
Common Names: Running pop, *bedoca*

Family: Passifloraceae (Passion flower)

Range: Native

Islands Inhabited: Floreana, Isabela (D,SN), San Cristóbal, Santa Cruz, Islet(s)

Habitat: Arid lowlands and moist uplands

Description: Herbaceous climbing vine with axillary tendrils, all vegetative parts oily and foul-smelling. *Leaves* alternate, simple; blade somewhat 3-lobed, ovate, 6–9 cm long, both surfaces covered with hairs, margins with glandular hairs. *Flowers* axillary, solitary, 4–5 cm across, with 3 highly dissected bracts beneath; calyx whitish, sepals 5; corolla white, petals 5; corona of numerous purple and white filaments. *Fruit* a berry, yellow, roundish, 2.5–3 cm long; seeds numerous, enclosed in a grayish mucilagenous substance.

Comments: This plant is extremely common and may be observed growing on trees, shrubs, rocks, and walls. The beautiful flowers open early in the morning and are usually visited by the Galápagos carpenter bee (*Xylocopa darwini*). By midmorning they close for good. Wiggins and Porter (1971) listed this plant as *P. foetida* var. *galapagensis* Killip.

347. *Passiflora foetida* (running pop), flower

Scientific Name: Cardiospermum galapageium B. L. Rob. & Greenm.

(Photo 348)

Common Name: Galápagos heartseed

Family: Sapindaceae (Soapberry)

Range: Endemic

Islands Inhabited: Isabela (SN), Santa Cruz, Santiago

Habitat: Arid lowlands

Description: Perennial vine, somewhat woody, climbing over other vegetation and rocks by means of tendrils, much-branched, stems grooved, somewhat hairy. *Leaves* alternate, compound, 2.5–9 cm long; leaflets 3, each of these with 3 parts, the terminal part larger than the 2 lateral parts; the middle leaflet to 7 cm long, its terminal part narrowly to broadly lanceolate, margins usually entire, occasionally with 2 lobes at the base. *Flowers* usually unisexual (plants monoecious), occasionally bisexual, in axillary clusters; corolla white, petals 4, 5 mm long, each with a yellow-tipped appendage, glandular-dotted; staminate flowers with 8 stamens. *Fruit* a capsule, brown when mature, obovoid, 2 cm long, somewhat papery and inflated, 3-angled; seeds 3, black, round, 4 mm in diameter.

Comments: Galápagos heartseed is considered rare. However, it can still be found in forested areas. The genus name is based on the Greek *kardia*, "heart," and *sperma*, "seed." This refers to a white, heart-shaped spot on each of the seeds.

348. *Cardiospermum galapageium* (Galápagos heartseed), flower

Scientific Name: Cardiospermum halicacabum L. (Photo 349)
Common Name: Heartseed

Family: Sapindaceae (Soapberry)

Range: Native; also known from other tropical regions throughout the world.

Islands Inhabited: Isabela (A), Pinzón, San Cristóbal, Santa Cruz, Santiago, Wolf

Habitat: Arid lowlands

Description: Perennial vine, somewhat woody, climbing over other vegetation and rocks by means of tendrils, much-branched, stems grooved, somewhat hairy. *Leaves* alternate, compound, 3.5–14 cm long; leaflets 3, each of these with 3 parts, the terminal part larger than the 2 lateral parts; the middle leaflet to 11.5 cm long, its terminal part ovate to broadly elliptic, margins serrate, dentate, or slightly lobed. *Flowers* usually unisexual (plants monoecious), occasionally bisexual, in axillary clusters; corolla white, petals 4, 4.5 mm long, each with a yellow-tipped appendage; staminate flowers with 8 stamens. *Fruit* a capsule, brown when mature, roundish, 2.5–3.5 cm long, somewhat papery and inflated, 3-angled; seeds 3, black, round, 3 mm in diameter.

Comments: This plant, listed as *C. corindum* L. in Wiggins and Porter (1971), is common in disturbed areas such as roadsides. Often it is grown as an ornamental for its flowers, or to cover trellises.

349. *Cardiospermum halicacabum* (heartseed), flowers and fruits

Scientific Name: Solandra maxima (Sessé & Moc.) P. S. Green
(Photo 350)
Common Names: Chalice vine, gold cup, *copa de oro*

Family: Solanaceae (Potato)

Range: Cultivated; also known from other regions of tropical America.

Islands Inhabited: Santa Cruz

Habitat: Arid lowlands

Description: Climbing, woody vine or shrub. *Leaves* alternate, simple, elliptic, to ca. 16 cm long, somewhat leathery, margins entire. *Flowers* axillary, solitary; corolla yellowish white to yellow, with several dark purple lines in the throat, cup-shaped with 5 lobes, to ca. 22 cm long; stamens 5. *Fruit* a berry, roundish; seeds numerous.

Comments: Chalice vine is often cultivated for its attractive, fragrant flowers. These persist for about four days, changing from yellowish white to a golden yellow. The genus was named for the Swedish naturalist and botanist Daniel C. Solander (1736–86).

350. *Solandra maxima* (chalice vine), flower

See also

Ipomoea imperati (p. 312)
I. triloba (p. 313)
Phaseolus adenanthus (p. 314)

Vines with Alternate Leaves and Yellow or Orange Flowers

Scientific Name: Merremia umbellata (L.) Hallier f. (Photo 351)
Common Name: Yellow merremia

Family: Convolvulaceae (Morning-glory)

Range: Introduced; also known from other tropical regions throughout the world.

Islands Inhabited: San Cristóbal, Santa Cruz

Habitat: Moist uplands

Description: Twining vine, young stems covered with hairs. *Leaves* alternate, simple, ovate, to ca. 14 cm long. *Flowers* in axillary umbels; corolla yellow, campanulate, ca. 3–3.5 cm long; stamens 5. *Fruit* a capsule, somewhat spherical, ca. 8 mm in diameter; seeds 4–6, dark brown.

Comments: This vine is common in disturbed areas and is often observed near towns and along roadsides. It is easily identifiable: According to van der Werff (1977), it is the only Galápagos member of the morning-glory family that has yellow flowers.

351. *Merremia umbellata* (yellow merremia), flowers

Scientific Name: Cucumis dipsaceus C. G. Ehrenb. ex Spach (Photo 352)
Common Names: Hedgehog gourd, teasel gourd, tiger's egg, *huevo de tigre*

Family: Cucurbitaceae (Gourd)

Range: Introduced; also known from other tropical regions throughout the world, originally from eastern Africa.

Islands Inhabited: Española, Floreana, Isabela (SN), San Cristóbal, Santa Cruz

Habitat: Arid lowlands

Description: Trailing annual vine to several m long, stems covered with bristly hairs. *Leaves* alternate, simple; blade cordate, reniform, or broadly ovate, 2–8 cm long, 3–9 cm wide, both surfaces covered with bristly hairs, margins minutely dentate; coiled tendrils unbranched. *Flowers* unisexual (plants monoecious), axillary. Staminate flowers solitary or in clusters of 2–4; corolla yellow, campanulate with 5 lobes, lobes 7–12 mm long; stamens 3. Pistillate flowers solitary or occasionally mixed with staminate flowers; corolla yellow, campanulate with 5 lobes, lobes 7–12 mm long; sterile stamens 3. *Fruit* a pepo, yellow, ellipsoid, 2–7 cm long, extremely spiny, pulp green; seeds numerous, pale brown.

Comments: Care must be taken when handling the fruits of this plant, as their spines can prove very painful. Hedgehog gourd is not used as a food source by humans, but *C. sativus* L. (cucumber) and *C. melo* L. (melon) are grown on Santa Cruz for their tasty fruits. The genus name, *Cucumis,* is Latin for "cucumber." Another member of this family, *Cucurbita pepo* L. (squash), is cultivated on Floreana and Isabela.

352. *Cucumis dipsaceus* (hedgehog gourd), flowers and fruits

Scientific Name: Mormordica charantia L. (Photo 353)
Common Names: Bitter melon, *achoccha silvestre*

Family: Cucurbitaceae (Gourd)

Range: Cultivated escape; also known from other tropical regions throughout the world, originally from Africa to Australia.

Islands Inhabited: Isabela (SN), Santa Cruz

Habitat: Arid lowlands

Description: Trailing or climbing annual vine to several m long. *Leaves* alternate, simple; blade broadly ovate to roundish, 5–12 cm long, 5- to 7-lobed, these occasionally somewhat secondarily lobed, margins dentate; coiled tendrils unbranched. *Flowers* unisexual (plants monoecious), axillary. Staminate flowers usually solitary; corolla yellow, petals 5, 1.5–2 cm long; stamens 3. Pistillate flowers usually solitary; corolla yellow, petals 5, 1.5–2 cm long. *Fruit* a pepo, orange tinged with red, ellipsoid to ovoid, 4–8 cm long, warty with longitudinal ribs, often opening into 3 sections; seeds numerous, tan, each covered with a bright red, fleshy outer coat.

Comments: This vine is extremely common and may be seen along roadsides and in most of the towns, often growing on rock walls and over other plants. Its flowers and fruits are quite noticeable, and the latter are often found torn open and missing their colorful seeds because birds have gotten to the fruits first and eaten the seeds. Wiggins and Porter (1971) mention that an extract of the roots in alcohol, taken in small amounts, is said to be an aphrodisiac. However, visitors should not attempt to test this hypothesis, as it is dangerous to taste unfamiliar plants without sufficient background knowledge.

353. *Mormordica charantia* (bitter melon), flowers and fruit

Scientific Name: Rhynchosia minima (L.) DC. (Photo 354)
Common Name: Rhynchosia

Family: Fabaceae (Pea)

Range: Native; also known from other tropical regions throughout the world.

Islands Inhabited: Española, Fernandina, Floreana, Genovesa, Isabela (A,CA,D,SN), Marchena, Pinta, Pinzón, Rábida, San Cristóbal, Santa Cruz, Santa Fe, Santiago

Habitat: Arid lowlands and moist uplands

Description: Perennial vine, slender stems herbaceous at first, becoming somewhat woody with age. *Leaves* alternate, odd-pinnately compound; leaflets 3, ovate to rhombic, to 8 cm long, to 7 cm wide, covered with glandular dots. *Flowers* in axillary racemes; corolla yellow, often with purplish or brownish lines, flower ca. 5–7 mm long, composed of 1 large standard petal (covered with glandular dots), 2 lateral wing petals, and 2 lower keel petals that are somewhat fused; stamens 10, 9 of these fused into a tube, the other free. *Fruit* a legume, ca. 1.5 cm long, covered with glandular dots; seeds 1–2.

Comments: This vine is typically found in disturbed sites such as roadsides and old fields.

354. *Rhynchosia minima* (rhynchosia), flowers

Scientific Name: Passiflora suberosa L. (Photo 355)
Common Name: Passion flower, *uvilla*

Family: Passifloraceae (Passion flower)

Range: Native; also known from other tropical regions throughout the world, originally from tropical America.

Islands Inhabited: Española, Fernandina, Floreana, Isabela (A,D), Pinta, San Cristóbal, Santa Cruz, Santiago, Islet(s)

Habitat: Moist uplands

Description: Herbaceous climbing vine with axillary tendrils, stems becoming angled. *Leaves* alternate, simple; blade usually 3-lobed but occasionally unlobed, lanceolate to ovate, 3–7 cm long. *Flowers* axillary, solitary or in pairs, 1–3 cm across; calyx greenish yellow, sepals 5; corolla absent; corona of numerous purple, yellow, and white filaments. *Fruit* a berry, dark purple, oblong to somewhat ovoid, to 1.5 cm long; seeds numerous, enclosed in a whitish mucilagenous substance.

Comments: This species tends to grow in rocky, shady areas. *Passiflora tridactylites* Hook. f., also native, is found on Española, Fernandina, Isabela (D), San Cristóbal, Santa Cruz, Santiago, and one or more of the smaller islands. It is quite similar in appearance to *P. suberosa*. However, it may be distinguished by the fact that its leaves have long, narrow lobes, and its androgynophore is extremely long (Lawesson 1988).

355. *Passiflora suberosa* (passion flower), flower

See also

Bougainvillea spectabilis (p. 299)
Solandra maxima (p. 306)

Vines with Alternate Leaves and Pink, Red, or Purple Flowers

Scientific Name: Ipomoea pes-caprae (L.) R. Br. (Photo 356)
Common Name: Beach morning-glory

Family: Convolvulaceae (Morning-glory)

Range: Native; also known from other tropical regions throughout the world.

Islands Inhabited: Genovesa, Isabela (CA,D,E,SN,W), Marchena, Santa Cruz, Islet(s)

Habitat: Coastal zone

Description: Perennial vine, stems to 10 m long, somewhat fleshy, containing a milky sap, long-trailing and often rooting at the nodes. *Leaves* alternate, simple; blade orbicular, broadly ovate, or elliptic, 5–10 cm long, somewhat leathery, margins entire. *Flowers* axillary, solitary or in cymes; corolla pinkish purple with a dark pinkish purple "star" extending from the throat, broadly funnelform, 2.5–3 cm long, ca. 2.5–3 cm across; stamens 5. *Fruit* a capsule, ovate, ca. 2 cm long; seeds to 4, dark brown, covered with brownish hairs.

Comments: The specific epithet of this plant, *pes-caprae*, refers to its leaves, which resemble the outline of a goat's foot. This vine is common along the archipelago's beaches, where it serves to help bind the sand. Its seeds are resistant to seawater and may travel for extended periods on the ocean's waves before being tossed up on another shore.

Ipomoea imperati (Vahl) Griseb. is a native that inhabits the beaches near Puerto Villamil (Isabela). This species, listed as *I. stolonifera* (Cyrill.) Gmel. in van der Werff (1977), is easily differentiated from *I. pes-caprae* by its white corolla with a yellow throat. Its current status is vulnerable.

356. *Ipomoea pes-caprae* (beach morning-glory), flower

Scientific Name: Ipomoea triloba L. (Photo 357)
Common Name: Morning-glory

Family: Convolvulaceae (Morning-glory)

Range: Native; also known from other tropical regions throughout the world, originally from tropical America.

Islands Inhabited: Baltra, Española, Fernandina, Floreana, Isabela (A,CA,D,E,SN), Pinta, Pinzón, San Cristóbal, Santa Cruz, Santiago, Wolf

Habitat: Arid lowlands

Description: Annual vine, stems to ca. 4 m long. *Leaves* alternate, simple; blade ovate to cordate, often 3-lobed, 4–10 cm long. *Flowers* axillary, solitary or in cymes; corolla pink or purplish with a dark purple throat, or occasionally white, funnelform, 1.5–2 cm long, 1–1.5 cm across; stamens 5. *Fruit* a capsule, somewhat spherical, 6–8 mm in diameter; seeds to 4, dark brown.

Comments: This is one of 11 species of *Ipomoea* in the archipelago. It is common in bright, open areas, as well as shaded forests, and is often seen in and around towns. The specific epithet, *triloba*, refers to the plant's leaves, which are often 3-lobed.

357. *Ipomoea triloba* (morning-glory), flowers and fruits

Scientific Name: Phaseolus mollis Hook. f. (Photo 358)
Common Name: Galápagos bean

Family: Fabaceae (Pea)

Range: Endemic

Islands Inhabited: Fernandina, Isabela (A,CA,D), Santa Cruz, Santiago

Habitat: Arid lowlands and moist uplands

Description: Herbaceous spreading vine to ca. 3 m long, young stems covered with hairs. *Leaves* alternate, odd-pinnately compound; leaflets 3, ovate, ca. 3–5.5 cm long. *Flowers* in racemes to ca. 30 cm long; corolla purplish with some white, composed of 1 large standard petal, 2 lateral wing petals, and 2 lower keel petals that are somewhat fused, the latter spiraled, flower ca. 1 cm long excluding the spiraled keel; stamens 10, 9 of these fused into a tube, the other free. *Fruit* a legume, oblong, 2.5–3 cm long, 6–7 mm wide; seeds 2–4.

Comments: Phaseolus adenanthus G. F. W. Mey, a native found on Española, also has oblong fruits. However, they are typically 7–12 cm long and 8–10 mm wide. This species also differs from *P. mollis* in having larger flowers (2–3 cm long) that are pink or purplish to white. In addition, its fruits are many-seeded.

Phaseolus atropurpureus DC., a native inhabiting Floreana, Isabela (A), Rábida, and Seymour, and *P. lathyroides* L., an introduced species on Floreana and San Cristóbal, produce linear-shaped fruits that are 2–3 mm wide. They differ from each other in that *P. atropurpureus* is low-growing and possesses a dark purple corolla, while *P. lathyroides* is relatively erect and has a dark red to dark purple corolla.

A cultivated member of this genus, *P. vulgaris* L., is found on San Cristóbal and Santa Cruz. It is known by a variety of common names, such as "kidney bean," "green bean," "snap bean," "wax bean," and *frijol.*

358. *Phaseolus mollis* (Galápagos bean), flowers

Scientific Name: Passiflora quadrangularis L. (Photo 359)
Common Names: Giant granadilla, *badea*

Family: Passifloraceae (Passion flower)

Range: Cultivated; also known from other regions of tropical America.

Islands Inhabited: Floreana, Isabela (SN), San Cristóbal, Santa Cruz

Habitat: Arid lowlands

Description: Climbing vine with axillary tendrils, stems 4-angled. *Leaves* alternate, simple; blade ovate to broadly elliptic-ovate, to 13 cm long, margins entire. *Flowers* axillary, solitary, 7–10 cm across; calyx reddish purple and white, sepals 5; corolla reddish purple and white, petals 5; corona of numerous purple and white filaments. *Fruit* a berry, green, oblong, 20–30 cm long; seeds numerous, enclosed in a whitish or grayish mucilagenous substance.

Comments: The fruit of giant granadilla is used for many purposes, including jellies, jams, and drinks.

359. *Passiflora quadrangularis* (giant granadilla), flower

See also

Bougainvillea spectabilis (p. 299)
Ipomoea linearifolia (p. 297)
I. nil (p. 316)

Vines with Alternate Leaves and Blue Flowers

Scientific Name: Ipomoea nil (L.) Roth (Photo 360)
Common Name: Blue morning-glory

Family: Convolvulaceae (Morning-glory)

Range: Introduced; also known from other tropical regions throughout the world.

Islands Inhabited: Floreana, Isabela (A), San Cristóbal, Santa Cruz, Santiago

Habitat: Arid lowlands

Description: Annual or perennial vine, stems to 5 m long, somewhat hairy. *Leaves* alternate, simple; blade broadly cordate to broadly ovate, usually somewhat 3-lobed, 3–18 cm long. *Flowers* axillary, solitary or in cymes; corolla blue or purplish with a white throat, funnelform, 2.5–6 cm long, 3–5 cm across; stamens 5. *Fruit* a capsule, ovoid, 7–8 mm long; seeds 3–6, dark brown or black.

Comments: Blue morning-glory is one of 11 species of *Ipomoea* found in the Galápagos.

360. *Ipomoea nil* (blue morning-glory), flowers

Vines with Alternate Leaves and Green Flowers

Scientific Name: Cissampelos pareira L. (Photo 361)
Common Name: Hairy cissampelos

Family: Menispermaceae (Moonseed)

Range: Native; also known from other tropical regions throughout the world.

Islands Inhabited: Fernandina, Floreana, Isabela (A,CA,D,SN), Pinta, Pinzón, San Cristóbal, Santa Cruz, Santiago

Habitat: Arid lowlands and moist uplands

Description: Climbing vine to 5 m long, sometimes longer, woody at the base, stems covered with short hairs. *Leaves* alternate, simple; blade broadly ovate to roundish, 3–11 cm long, 2.5–10 cm wide, surfaces covered with soft hairs, margins entire. *Flowers* unisexual (plants monoecious). Staminate flowers in axillary clusters; corolla greenish, dishlike with a small rim, 1–1.4 mm across; stamens 4, fused into a single column. Pistillate flowers in axillary clusters; corolla greenish, petals 2, occasionally somewhat fused, ca. 1 mm long. *Fruit* a drupe, strongly recurved, 4–5 mm long, somewhat hairy; seed 1.

Comments: Cissampelos glaberrima A. St.-Hil. (smooth cissampelos), listed as *C. galapagensis* Stewart in Wiggins and Porter (1971), is differentiated by having smooth stems and leaves, smooth fruits, and a whitish covering on the lower surface of the leaves. It is an endemic and is known only from Santa Cruz and Santiago. Both of these species appear to prefer forested, somewhat shaded areas.

361. *Cissampelos pareira* (hairy cissampelos), flowers

Vines With Opposite Leaves and Yellow or Orange Flowers

Scientific Name: Allamanda cathartica L. (Photo 362)
Common Names: Golden trumpet, yellow allamanda

Family: Apocynaceae (Dogbane)

Range: Cultivated; also known from other regions of tropical America, originally from northern South America.

Islands Inhabited: San Cristóbal, Santa Cruz

Habitat: Arid lowlands and moist uplands

Description: Perennial vine to ca. 15 m long, sometimes shrublike, stems containing a milky sap. *Leaves* opposite or in whorls of 3–4, simple, oblanceolate to elliptic-oblong, 10–15 cm long, margins entire. *Flowers* in cymes; corolla yellow with brownish yellow stripes inside the throat, funnelform with 5 lobes, to 12 cm across; stamens 5. *Fruit* a capsule, roundish, to 4 cm across, spiny; seeds numerous.

Comments: This plant's attractive flowers are quite noticeable. Its primary use is as an ornamental, and it is often seen covering walls and fences or clinging to the sides of houses. On the mainland, however, this plant is also used in the preparation of a laxative.

362. *Allamanda cathartica* (golden trumpet), flowers

Vines with Opposite Leaves and Pink, Red, or Purple Flowers

Scientific Name: Sarcostemma angustissimum (Andersson) R. W. Holm (Photo 363)

Common Name: Galápagos sarcostemma

Family: Asclepiadaceae (Milkweed)

Range: Endemic

Islands Inhabited: Baltra, Española, Fernandina, Floreana, Isabela (A,CA,D,E,SN,W), Pinta, Pinzón, Rábida, San Cristóbal, Santa Cruz, Santiago, Islet(s)

Habitat: Arid lowlands

Description: Vinelike perennial, slender, branching, spreading over other vegetation and rocks. *Leaves* opposite, simple, linear to narrowly oblong, 1–6 cm long, margins entire, slightly rolled under. *Flowers* in terminal, umbel-like cymes; corolla dark purplish, 5-lobed, lobes 2.5–3 mm long; corona yellowish white; stamens 5. *Fruit* a follicle, spindle-shaped, 8–12 cm long; seeds numerous, brown, flask-shaped, each with a tuft of silky white hairs.

Comments: This plant, listed as *S. angustissima* (Andersson) R. W. Holm in Wiggins and Porter (1971), easily spreads by means of its wind-borne seeds. It is common on the lava fields near the town of Puerto Villamil (Isabela).

363. *Sarcostemma angustissimum* (Galápagos sarcostemma), flowers

Vines with Whorled Leaves and Yellow or Orange Flowers

See:

Allamanda cathartica (p. 318)

Cacti with White Flowers

Scientific Name: Brachycereus nesioticus (K. Schum.) Backbg. (Photos 364–65)
Common Names: Lava cactus, *cactillo de lava*

Family: Cactaceae (Cactus)

Range: Endemic

Islands Inhabited: Fernandina, Genovesa, Isabela (CA), Pinta, Santiago, Islet(s)

Habitat: Arid lowlands

Description: Fleshy perennial, stems cylindrical and arranged in clumps, each stem 10–50 cm tall and with 16–22 ribs, yellow, greenish yellow, or brownish yellow. *Leaves* represented by spines, ca. 40 per cluster, white, yellow, or black, to 5 cm long. *Flowers* solitary, 6–11 cm long, 2–5.5 cm across; inner perianth parts white to yellowish white, 10–18 mm long; stamens numerous. *Fruit* a berry, red to reddish brown, roundish, 1.5–3.5 cm long, 1–1.4 cm across, fleshy, covered with yellow spines; seeds numerous, brownish black.

Comments: Brachycereus is one of seven angiosperm genera endemic to the Galápagos. The genus name of this rare plant is derived from the Greek *brachys,* "short," and the Latin *cereus,* "candle," both of which refer to this cactus's unmistakable shape. The common name alludes to this plant's ability to thrive on barren lava flows, both pahoehoe and aa. The fact that this genus includes only a single species suggests that it has not inhabited the islands for as long as *Opuntia,* which has speciated extensively since arriving.

364. (above) *Brachycereus nesioticus* (lava cactus)

365. (right) *Brachycereus nesioticus* (lava cactus), fruits

Scientific Name: Hylocereus undatus (Haw.) Britton & Rose (Photo 366)
Common Names: Night-blooming cereus, *flor de Cáliz*

Family: Cactaceae (Cactus)

Range: Cultivated; also known from other regions of tropical America.

Islands Inhabited: Santa Cruz

Habitat: Arid lowlands

Description: Climbing terrestrial or epiphytic perennial, stems fleshy, 3-winged, wings 2–3 cm wide, margins wavy, green. *Leaves* represented by spines, 1–4 per cluster, 1–3 mm long. *Flowers* solitary, 25–30 cm long, 15–25 cm across; inner perianth parts white, 10–15 cm long; stamens numerous. *Fruit* a berry, red, oblong, 5–12.5 cm long, fleshy; seeds numerous, shiny black.

Comments: The fragrant flowers of night-blooming cereus open in the evening and last only until midmorning of the next day, when they close and wilt. Early in the morning, one is likely to see these beautiful flowers being visited by the Galápagos carpenter bee (*Xylocopa darwini*), its back covered with yellow pollen from the abundant stamens. Although able to produce seeds, this plant more commonly spreads by vegetative means. As it climbs, its stems often produce aerial roots.

The amateur botanist might be tempted to mistake this plant for *Euphorbia lactea* Haw. (Photo 367), a cultivated member of the Euphorbiacae that also has winged stems and small spines. However, the latter plant, known as the mottled spurge, can grow much taller and has white longitudinal stripes on its stems. In addition, it possesses a milky sap, and its flowers are much smaller and quite different than those of a cactus. It can be seen on Santa Cruz.

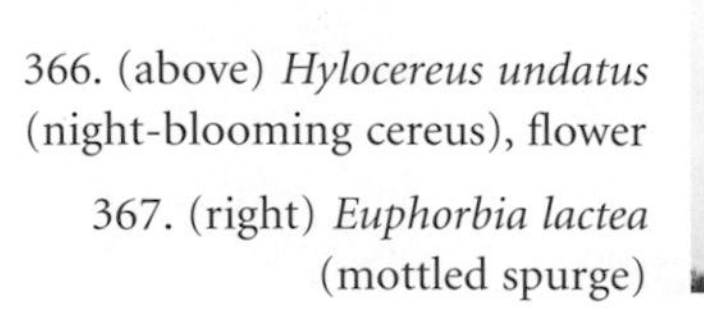

366. (above) *Hylocereus undatus* (night-blooming cereus), flower

367. (right) *Euphorbia lactea* (mottled spurge)

Cacti with Yellow or Orange Flowers

Scientific Name: Jasminocereus thouarsii (Weber) Backbg. (Photos 368–71)
Common Names: Candelabra cactus, *candelabro, cacto esbelto, cirio*

Family: Cactaceae (Cactus)

Range: Endemic

Islands Inhabited: Fernandina, Floreana, Isabela (A,CA,D,E,SN,W), San Cristóbal, Santa Cruz, Santiago, Islet(s)

Habitat: Arid lowlands

Description: Fleshy perennial to 7 m tall, trunk and branches with 11–22 ribs, green to greenish yellow. *Leaves* represented by spines, up to 35 per cluster, white, yellow, reddish brown or black, to 9 cm long. *Flowers* solitary, 5–9 cm long, 2–6 cm across; inner perianth parts yellow to yellowish green, to 2.5–4 cm long; stamens numerous. *Fruit* a berry, greenish to reddish purple, somewhat roundish to oblong, 1.5–7 cm long, 1.5–4.5 cm across, fleshy; seeds numerous, black.

Comments: Jasminocereus, one of seven angiosperm genera endemic to the Galápagos, consists of a single species with three varieties. *Jasminocereus thouarsii* var. *sclerocarpus* (K. Schum.) Anderson & Walkington (Photo 368) occurs only on Fernandina and Isabela (A,CA,D,E,SN,W). It has 10–35 spines per cluster, waxy-textured flowers, and greenish red fruits that are 4–7 cm long and 3–4.5 cm across.

Jasminocereus thouarsii var. *thouarsii* possesses 8–21 spines per cluster, flowers that are not waxy, and green fruits that are 2.8–5 cm long and 1.8–3 cm across. It is found on Floreana, San Cristóbal, and the islet known as Corona del Diablo.

Jasminocereus thouarsii var. *delicatus* (Dawson) Anderson & Walkington (Photos 369–71) has 10–22 spines per cluster, flowers that are not waxy, and reddish purple fruits that are 1.5–4.4 cm long and 3.5–4.2 cm across. This is the most familiar member to most visitors, since it inhabits Santa Cruz and occurs along the road from Puerto Ayora to the CDRS. It is also found on Santiago and on one or more of the smaller islands. Occasionally the fruits are used to make a refreshing juice. Although all of these varieties are considered rare, they are still quite easy to find.

368. (left) *Jasminocereus thouarsii* var. *sclerocarpus* (candelabra cactus)

369. (above) *Jasminocereus thouarsii* var. *delicatus* (candelabra cactus)

370. (above) *Jasminocereus thouarsii* var. *delicatus* (candelabra cactus), flower

371. (right) *Jasminocereus thouarsii* var. *delicatus* (candelabra cactus), fruit

Scientific Name: Opuntia echios J. T. Howell (Photos 372–77; see also Photo 47)
Common Names: Prickly pear cactus, *tuna*

Family: Cactaceae (Cactus)

Range: Endemic

Islands Inhabited: Baltra, Daphne, Isabela (SN), Santa Cruz, Santa Fe, Seymour, Islet(s)

Habitat: Arid lowlands and moist uplands

Description: Shrubby to treelike perennial to 12 m tall, branches often hanging downward. Trunk, when present, somewhat woody, to 1.25 m across, covered with spines when young, these persisting or later replaced with reddish orange "bark." Stem pads fleshy, to 50 cm long and 32 cm wide. *Leaves* 2–6 mm long, nonpersistent, usually not obvious. Spines 1–50 per cluster, yellow, brownish, or white, 1–12 cm long; glochids present to absent. *Flowers* solitary, 5.2–10 cm long, ca. 5–9 cm across; inner perianth parts yellow, 2.6–4.7 cm long; stamens numerous. *Fruit* a berry, green to yellowish green or brownish, ovate, roundish, or top-shaped, 4–11.7 cm long, 2.8–4.7 cm across, fleshy, spines and glochids present to few; seeds numerous, whitish brown, 2–5 mm long.

Comments: This species includes five varieties. Technically, each is classified as rare, but they are still relatively easy to find. *Opuntia echios* var. *barringtonensis* Dawson (Photo 372), the only prickly pear cactus inhabiting Santa Fe, derives its specific epithet from the island's English name, Barrington, honoring the British admiral S. Barrington. It does not occur on any other islands, and it differs from the other four varieties in having fruits that typically are more than 7 cm long when mature. These are truly impressive cacti, with trunks up to 1.25 m in diameter. Appearing like sentinels on the cliffs, they are one of the lasting memories for most visitors to Santa Fe.

Opuntia echios var. *echios* (Photo 373) is the only prickly pear cactus inhabiting Baltra and Daphne. It also occupies Santa Cruz and one or more of the smaller islands. Treelike specimens dominate the landscape on Plaza Sur, while shrubby forms may be seen on Daphne and the northern slope of Santa Cruz. This plant may reach a height of 3 m, has top-shaped fruits, and produces seeds 2–3 mm long. Ground finches (*Geospiza* spp.) have been observed feeding on its pollen, nectar, seeds, and arils on Daphne (Grant and Grant 1981).

Opuntia echios var. *gigantea* (J. T. Howell) D. M. Porter (Photos 374–76) occurs only on Santa Cruz. This variety of prickly pear cactus includes the tallest members of the genus (3–12 m). One explanation for the extraordinary height of these cacti is that selection favored those plants that produced pads that were out of reach of hungry tortoises. In fact, a correlation does seem to exist between cactus form and tortoise range. Those islands that now have tortoises, or that were inhabited by tortoises in the past, possess the treelike cacti. Islands without a history of tortoises demonstrate the shrubby form. This plant, which produces roundish to oblong fruits, is often seen growing together with *Jasminocereus thouarsii* var. *delicatus* (Cactaceae) on Santa Cruz. Remember that the prickly pear cactus has pads, while the candelabra cactus does not. Finches (*Geospiza* spp.) are known to feed on the pollen, nectar, seeds, and arils of this cactus, and mockingbirds (*Nesomimus parvulus*) take the nectar (Grant and Grant 1981). In addition, ants feed on an extrafloral nectar that is produced at the base of this cactus's young spines (Meier 1993).

Opuntia echios var. *inermis* Dawson (Photo 377) is known only from Volcán Sierra Negra (Isabela) and is easily found on the lava fields near the town of Puerto Villamil. This treelike cactus may reach a height of 6 m and produces roundish fruits 4–5.2 cm long. The only other prickly pear cactus on Sierra Negra is *O. insularis* (see Photo 381). It differs in that it is shorter (1–2.5 m) and typically produces smaller fruits (2–4.2 cm long).

Opuntia echios var. *zacana* (J. T. Howell) Anderson & Walkington (see Photo 47) is the only prickly pear cactus found on Seymour, and it occurs on no other islands. Typically it takes on a shrubby or low-growing form (1–2 m tall) and never becomes truly treelike. It has top-shaped fruits and produces seeds 3–4 mm long. Visitors may notice that it frequently grows next to *Bursera malacophylla* (Galápagos incense tree).

372. *Opuntia echios* var. *barringtonensis* (prickly pear cactus)

373. *Opuntia echios* var. *echios* (prickly pear cactus)

374. (left) *Opuntia echios* var. *gigantea* (prickly pear cactus)

375. (above) *Opuntia echios* var. *gigantea* (prickly pear cactus), flowers

376. (above) *Opuntia echios* var. *gigantea* (prickly pear cactus), flower

377. (right) *Opuntia echios* var. *inermis* (prickly pear cactus) and *Jasminocereus thouarsii* var. *sclerocarpus* (candelabra cactus)

Scientific Name: Opuntia galapageia Hensl. (Photos 378–79)
Common Names: Prickly pear cactus, *tuna*

Family: Cactaceae (Cactus)

Range: Endemic

Islands Inhabited: Pinta, Pinzón, Rábida, Santiago, Islet(s)

Habitat: Arid lowlands

Description: Low-growing to treelike perennial to 5 m tall. Trunk, when present, somewhat woody, covered with spines when young, later covered with reddish to reddish black "bark." Stem pads fleshy, to 38 cm long and 27 cm wide. *Leaves* 2–9 mm long, nonpersistent, usually not obvious. Spines 5–35 per cluster, yellow, orange-red, brown, or white, to 7.5 cm long; glochids few or absent. *Flowers* solitary, 4–7 cm long, 3.5–6 cm across; inner perianth parts yellow, 1.7–4 cm long; stamens numerous. *Fruit* a berry, green to yellowish green or brownish, roundish to oblong, 1.7–6 cm long, 2–4 cm across, fleshy, spines and glochids few or absent; seeds numerous, light brown, 2–5 mm long.

Comments: This species includes three varieties, all of which are considered rare. *Opuntia galapageia* var. *galapageia* (Photo 378) occurs on Pinta, Santiago, and one or more of the smaller islands. Grant and Grant (1981) observed that ground finches (*Geospiza* spp.) on Pinta feed on this plant's pollen, nectar, and seeds, while mockingbirds (*Nesomimus parvulus*) eat the pollen and nectar.

Opuntia galapageia var. *macrocarpa* Dawson is easy to identify, as it is the only prickly pear cactus found on Pinzón, and it occurs nowhere else. Likewise, *O. galapageia* var. *profusa* Anderson & Walkington (Photo 379) is the only prickly pear cactus that grows on Rábida, and it is found on no other islands. Its variety name comes from the fact that it produces an extraordinary number of fruits on each pad (up to 82).

378. *Opuntia galapageia* var. *galapageia* (prickly pear cactus)

379. *Opuntia galapageia* var. *profusa* (prickly pear cactus) and Galápagos mockingbird (*Nesomimus parvulus*)

Scientific Name: Opuntia helleri K. Schum. (Photo 380)
Common Names: Prickly pear cactus, *tuna*

Family: Cactaceae (Cactus)

Range: Endemic

Islands Inhabited: Darwin, Genovesa, Marchena, Wolf

Habitat: Arid lowlands

Description: Low-growing perennial, often forming large clumps, occasionally shrub or small tree to ca. 2 m tall. Trunk, when present, somewhat woody, covered with reddish brown "bark." Stem pads fleshy, to 37 cm long and 22 cm wide. *Leaves* 0.4–6 mm long, nonpersistent, usually not obvious. Spines 7–28 per cluster, yellowish white or brown, to ca. 5 cm long, often wavy and somewhat flexible; glochids usually present. *Flowers* solitary, 4–8 cm long, 3–5.5 cm across; inner perianth parts yellow, 2.5–3.5 cm long; stamens numerous. *Fruit* a berry, green, roundish to oblong, 4–7 cm long, 2–4 cm across, fleshy, spines and glochids present; seeds numerous, whitish yellow, 4–6 mm long.

Comments: This is the only prickly pear cactus inhabiting Genovesa. Studies by Grant and Grant (1981) indicate that ground finches (*Geospiza* spp.) feed on the pollen, nectar, seeds, and arils of this rare plant. Mockingbirds (*Nesomimus parvulus*) eat the pollen, nectar, and arils, and doves (*Zenaida galapagoensis*) consume the pollen.

380. *Opuntia helleri* (prickly pear cactus) and *Bursera graveolens* (incense tree)

Scientific Name: Opuntia insularis A. Stewart (Photo 381)
Common Names: Prickly pear cactus, *tuna*

Family: Cactaceae (Cactus)

Range: Endemic

Islands Inhabited: Fernandina, Isabela (A,CA,D,SN,W)

Habitat: Arid lowlands and moist uplands

Description: Shrubby to treelike perennial to 2.5 m tall. Trunk, when present, somewhat woody, covered with spines when young, later covered with reddish orange "bark." Stem pads fleshy, to 52 cm long and 25 cm wide. *Leaves* 5–8 mm long, nonpersistent, usually not obvious. Spines 10–50 per cluster, yellowish, reddish, or brown, to 5 cm long; glochids usually present. *Flowers* solitary; inner perianth parts yellow, 0.8–1.5 cm long; stamens numerous. *Fruit* a berry, greenish, roundish, 2–4.2 cm long, 2–3 cm across, fleshy, spines and glochids present; seeds numerous, whitish brown, 2.5–3.5 mm long.

Comments: This species, considered rare, is the only prickly pear cactus that occurs on Fernandina. It is also the only prickly pear cactus found on Volcán Alcedo, Volcán Darwin, and Volcán Wolf of Isabela. However, it shares Volcán Cerro Azul with *Opuntia saxicola* J. T. Howell. *Opuntia saxicola,* also considered rare, occurs nowhere else in the archipelago. It differs from *O. insularis* in having spines that typically are more than 5 cm in length (to 8 cm). In addition, its inner perianth parts are 2–2.5 cm long, and it produces top-shaped fruits.

Finally, *O. insularis* shares Volcán Sierra Negra with *O. echios* var. *inermis* (see Photo 377). The latter differs from *O. insularis* in that it is taller (to 6 m) and usually produces larger fruits (4–5.2 cm long).

381. *Opuntia insularis* (prickly pear cactus), flower and fruits

Scientific Name: Opuntia megasperma J. T. Howell (Photos 382–83)
Common Names: Prickly pear cactus, *tuna*

Family: Cactaceae (Cactus)

Range: Endemic

Islands Inhabited: Española, Floreana, San Cristóbal, Islet(s)

Habitat: Arid lowlands and moist uplands

Description: Shrubby to treelike perennial to 6 m tall, branches typically not hanging downward. Trunk, somewhat woody, to 1 m across, covered with spines when young, later covered with reddish or brownish "bark." Stem pads fleshy, to 48 cm long and 35 cm wide. *Leaves* 3–11 mm long, nonpersistent, usually not obvious. Spines 1–50 per cluster, yellow, brownish or blackish, 2.5–10 cm long; glochids few or absent. *Flowers* solitary, 6–13 cm long, 6–10.5 cm across; inner perianth parts yellow to reddish yellow, 2.5–5 cm long; stamens numerous. *Fruit* a berry, green to yellowish green, top-shaped, 4–17 cm long, 2.7–5.3 cm across, fleshy, spines present; seeds numerous, light brown, 5–13 mm long.

Comments: This species includes three varieties. *Opuntia megasperma* var. *megasperma* (Photo 382), considered vulnerable, inhabits Corona del Diablo and Floreana. In fact, it is the only prickly pear cactus inhabiting these islands. It produces larger spines (6–10 cm long), flowers (11–13 cm long), and fruits (8–17 cm long) than either of the other two varieties. The specific epithet and variety name come from the Greek *mega*, "large," and *spermus*, "seed," because it produces the largest seeds (8–13 mm long) of all the prickly pear cacti in the archipelago.

Opuntia megasperma var. *orientalis* (J. T. Howell) D. M. Porter (Photo 383) takes its variety name from the Latin word for "eastern." This is because the plant inhabits the two easternmost islands in the archipelago, Española and San Cristóbal. It also inhabits one or more of the smaller islands. This variety, considered rare, includes members whose trunks are among the thickest in the archipelago (to 60 cm across). It differs from *O. megasperma* var. *megasperma* in having smaller spines (2.9–5.5 cm long), flowers (8.5–11 cm long), and fruits (6–13 cm long). The pollen, seeds, and arils of *Opuntia megasperma* var. *orientalis* on Española serve as a food source for certain ground finches (*Geospiza* spp.). In addition, mockingbirds (*Nesomimus macdonaldi*) feed on the pollen and arils, while doves (*Zenaida galapagoensis*) concentrate on the arils (Grant and Grant 1981).

Opuntia megasperma var. *mesophytica* J. Lundh., also considered rare, is found on San Cristóbal and one or more of the smaller islands. It differs from *O. megasperma* var. *orientalis* primarily in having a thinner trunk (to 40 cm across), shorter spines (2.5–3.9 cm), shorter flowers (6–8 cm), and shorter fruits (4–6 cm). In addition, *O. megasperma* var. *mesophytica* normally grows at higher elevations.

382. *Opuntia megasperma* var. *megasperma* (prickly pear cactus), fruits

383. *Opuntia megasperma* var. *orientalis* (prickly pear cactus)

Glossary

Aa lava Lava with a rough, broken appearance; one of two major types in the Galápagos Islands.

Achene A dry, indehiscent, single-seeded fruit formed from a single carpel; the seed is attached to the surrounding fruit wall at only one point.

Aggregrate fruit A fruit formed from a single flower with many separate carpels.

Alternate Borne singly at each node, as leaves on a stem.

alternate

Androgynophore A stalk that bears both stamens and pistil; typically rises above the other parts a flower.

Angiosperm A plant that produces flowers, and seeds inside a fruit.

Annual A plant that completes its life cycle and dies within one growing season.

Anther The pollen-producing part of a stamen, usually borne at the top of the filament.

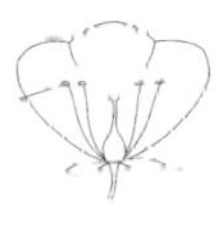

anther

Apex The tip of a structure.

Archipelago A large group of islands.

Aril A fleshy covering over all or part of a seed.

Awn A bristlelike appendage.

Axillary Located in the upper angle formed between a leaf or branch and the stem.

Basal Produced at the base of a plant, as leaves on a stem.

basal

Beak A slender point on the tip of a fruit.

Berry A fleshy, indehiscent, one- to many-seeded fruit formed from a single carpel or two or more fused carpels.

Bipinnately compound A compound leaf that is twice pinnate; each of the major divisions is pinnately divided.

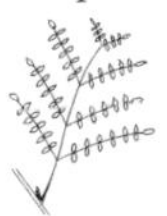

bipinnately compound

Bisexual Having both functional stamens and pistil(s) in each flower; perfect.

Blade The expanded part of a leaf.

Bract A modified leaf, often scalelike and associated with flowers.

Bracteole A secondary or very small bract.

Buttress A flared, basal part of a tree trunk that functions in support.

Ca. Approximately.

Calyx A collective term for all of the sepals of a flower.

Campanulate Bell-shaped.

Capsule A dry, dehiscent, many-seeded fruit formed from two or more fused carpels.

Carpel The part or parts of a flower that bear the ovules; see pistil.

CDF Charles Darwin Foundation.

CDRS Charles Darwin Research Station.

clustered

Clustered Grouped closely together, as leaves on a stem.

Cm Centimeter (1 cm = 0.394 inch).

Compound leaf A leaf composed of two or more leaflets.

cordate

Cordate Heart-shaped, with the notch at the base; refers to the overall shape of a structure or just the base.

Corolla A collective term for all of the petals of a flower.

Corona Appendages, often petal-like, that are located between the corolla and stamens in some flowers.

corymb

Corymb A general category of inflorescence in which the lower or outermost flowers open first; usually broad and more or less flat-topped.

crenate

Crenate Having shallow, rounded teeth; scalloped.

Cultivated Intentionally brought into one area (from another) by humans for the purpose of cultivation.

Cultivated escape A cultivated plant that has also become established outside of cultivation.

Cyathium An inflorescence that consists of a cup-shaped structure containing several male flowers (each with one stamen) and/or a single female flower (composed of three fused carpels). The cup-shaped structure often possesses glands and/or small petal-like appendages. This inflorescence is typical of many members of the Euphorbiaceae.

cyme

Cyme A general category of inflorescence in which the upper or innermost flowers open first; usually broad and more or less flat-topped or round-topped.

Deciduous Dropping leaves during a specific season.

Dehiscent Opening at maturity, releasing its contents.

dentate

Dentate Having sharp, spreading teeth that are arranged more or less perpendicular to the margin.

Determinate Of an inflorescence, in which the upper or innermost flowers open first.

Dioecious Having staminate and pistillate flowers borne on separate individuals of a species.

disc flower

Disc flower A type of flower, found in many members of the Asteraceae, that has a tubular corolla, typically with five lobes.

Discoid head Composed entirely of disc flowers.

Dissected Deeply divided into many narrow sections.

Drupaceous Outwardly resembling a drupe in general appearance.

Drupe A fleshy, indehiscent, one-seeded fruit formed from a single carpel, with a hard inner layer surrounding the seed.

Ellipsoid Elliptic in long section and circular in cross section, in reference to a solid body.

Elliptic In the shape of a narrow oval; widest at the middle and more or less equally narrowed toward both ends.

elliptic

Endemic Restricted to a particular geographic region. Plants that are endemic to the Galápagos Islands are found nowhere else in the world.

Entire Having a continuous margin; lacking teeth or divisions.

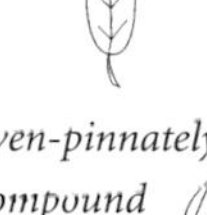

Epiphyte A plant that typically grows on another plant but gains no nutrients from it.

Even-pinnately compound Describes a pinnately compound leaf with an even number of leaflets, lacking a single, terminal leaflet.

even-pinnately compound

Exotic Any organism brought into one area from another by humans.

Feral Having returned to the wild state after being once domesticated.

Fibrous Resembling fibers.

Filament The stalk of a stamen; an anther is borne at the top.

filament

Filiform Long and slender, more or less threadlike.

Floret A small flower found within the spikelet of a grass inflorescence.

Floristics The study of the plants of a particular region.

Flower The reproductive structure of an angiosperm, typically consisting of stamens and/or pistil(s) and a perianth.

filiform

Follicle A dry, dehiscent, many-seeded fruit formed from a single carpel and opening along a single side.

flower

Fruit The seed-bearing structure of a flowering plant; the ripened ovary.

Funnelform Funnel-shaped.

Garúa A mixture of light rain and/or mist; typical of the Galápagos cool season.

Gland A protuberance, appendage, or other structure that typically produces a sticky or oily substance.

Glandular Having glands.

Glochid A tiny barbed bristle, often growing in clumps at the base of cactus spines.

Glume One of two bracts at the base of a grass spikelet.

GNPS Galápagos National Park Service.

Head A dense, compact inflorescence of sessile or nearly sessile flowers arranged on a very short axis.

head

Herb A plant that lacks a persistent, above-ground, woody stem.

Hesperidium A fleshy, indehiscent, many-seeded fruit formed from several fused carpels, with a leathery, oil-containing skin.

Imperfect flower Having either stamens or pistil(s), but not both; unisexual.
Impressed Pressed inward, as with leaf veins on certain leaves.
Indehiscent Remaining closed at maturity.
Indeterminate Of an inflorescence, in which the lower or outermost flowers open first.
Inflorescence The arrangement of the flowers on a floral axis.
Introduced Unintentionally brought into one area from another by humans.

Keel The two partly fused lower petals found in the flower of many members of the Fabaceae.
Kilometer 1,000 meters, equal to ca. 0.621 mile.

Labellum The lower, typically enlarged petal of an orchid flower.
lanceolate
Lanceolate Lance-shaped; much longer than broad, widest below the middle and tapering toward the apex.
Lateral At or on the side.
Leaflet One of the segments of a compound leaf.
Leaf scar A scar on a stem that indicates where a leaf was attached.
Legume A dry, dehiscent, many-seeded fruit formed from a single carpel and opening along two sides.
Lemma The larger of two bracts found at the base of a grass floret.
Lenticel A spongy area in the bark of a plant that allows the exchange of gases between the atmosphere and the inside of the plant; typically appearing as small dots or lines on the stem.
Ligulate head Composed entirely of ray flowers.
Limb The expanded part of a sympetalous corolla.
linear
Linear Long and narrow, with nearly parallel sides.
Lobe A section or division of a plant part, such as a leaf or corolla.
Loment A dry, deciduous, many-seeded fruit formed from a single carpel and opening along two sides; constrictions between the seeds form many one-seeded sections.

M Meter (1 m = 39.37 inches).
Margin The outer edge of a structure, such as a leaf blade.
Mm Millimeter (1 mm = 0.039 inch).
Monoecious Having staminate and pistillate flowers borne on each individual of a species.
Multiple fruit A fruit formed from many flowers, each with a single carpel, crowded together on a single axis.

Native Occurring naturally in a particular region. Plants that are native to the Galápagos Islands are found there naturally, as well as in other parts of the world.

Nerve A conspicuous, simple, unbranched vein or rib of a leaf or other structure.

Node The point on a stem where leaves or branches typically arise; occasionally roots also form here.

Nut A dry, indehiscent, one-seeded fruit formed from a single carpel or two or more fused carpels; the surrounding fruit wall is hard or bony.

Nutlet A small nut; a dry, indehiscent, typically one-seeded fruit segment of some members of the Boraginaceae, Lamiaceae, and Verbenaceae.

Obcordate Heart-shaped, with the notch at the apex.

obcordate

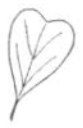

Oblanceolate Lance-shaped, with the broadest point above the middle and tapering toward the base.

oblanceolate

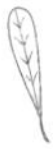

Oblong Much longer than broad, with nearly parallel sides.

oblong

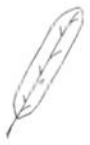

Obovate Egg-shaped, with the broadest point toward the apex.

obovate

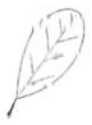

Obtuse Blunt or rounded at the apex.

Odd pinnately compound Describes a pinnately compound leaf with an odd number of leaflets and a single, terminal leaflet.

odd-pinnately compound

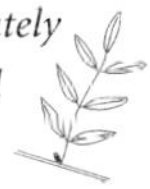

Opposite Borne in pairs at each node, across from each other, as leaves on a stem.

opposite

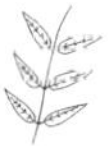

Orbicular More or less circular in outline.

orbicular

Ovary The enlarged, basal portion of the pistil that contains the ovules.

ovary

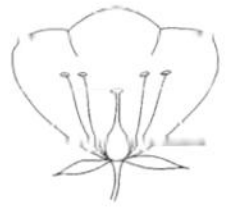

Ovate Egg-shaped, with the broadest point toward the base.

ovate

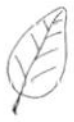

Ovoid Egg-shaped, in reference to a solid.

Ovule An immature seed.

Pahoehoe lava Lava with a smooth, somewhat ropy appearance; one of two major lava types in the Galápagos Islands.

Palea The smaller of two bracts found at the base of a grass floret.

Palmately compound Describes a compound leaf with the leaflets arising from a common point.

palmately compound

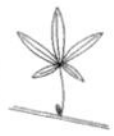

Panicle An indeterminate, branching inflorescence, the branches typically racemes or corymbs.

panicle

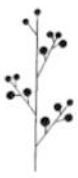

Pepo A fleshy, indehiscent, many-seeded fruit formed from a single carpel, with a relatively thick, leathery skin.

Perennial Living more than two years.

Perfect flower Having both stamens and pistil(s); bisexual.

Perianth A collective term for the calyx and/or corolla.

Petal One part of the corolla of a flower, usually white or variously colored.

petal

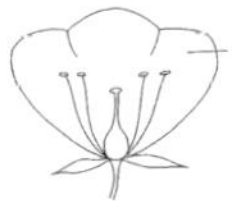

Petiole The stalk of a leaf.

Phloem The conducting tissue that transports food throughout a plant's body.

Pinna One of the primary divisions of a pinnately compound leaf.

Pinnately compound Describes a compound leaf with the leaflets arranged along the entire length of an axis; there may or may not be one terminal leaflet.

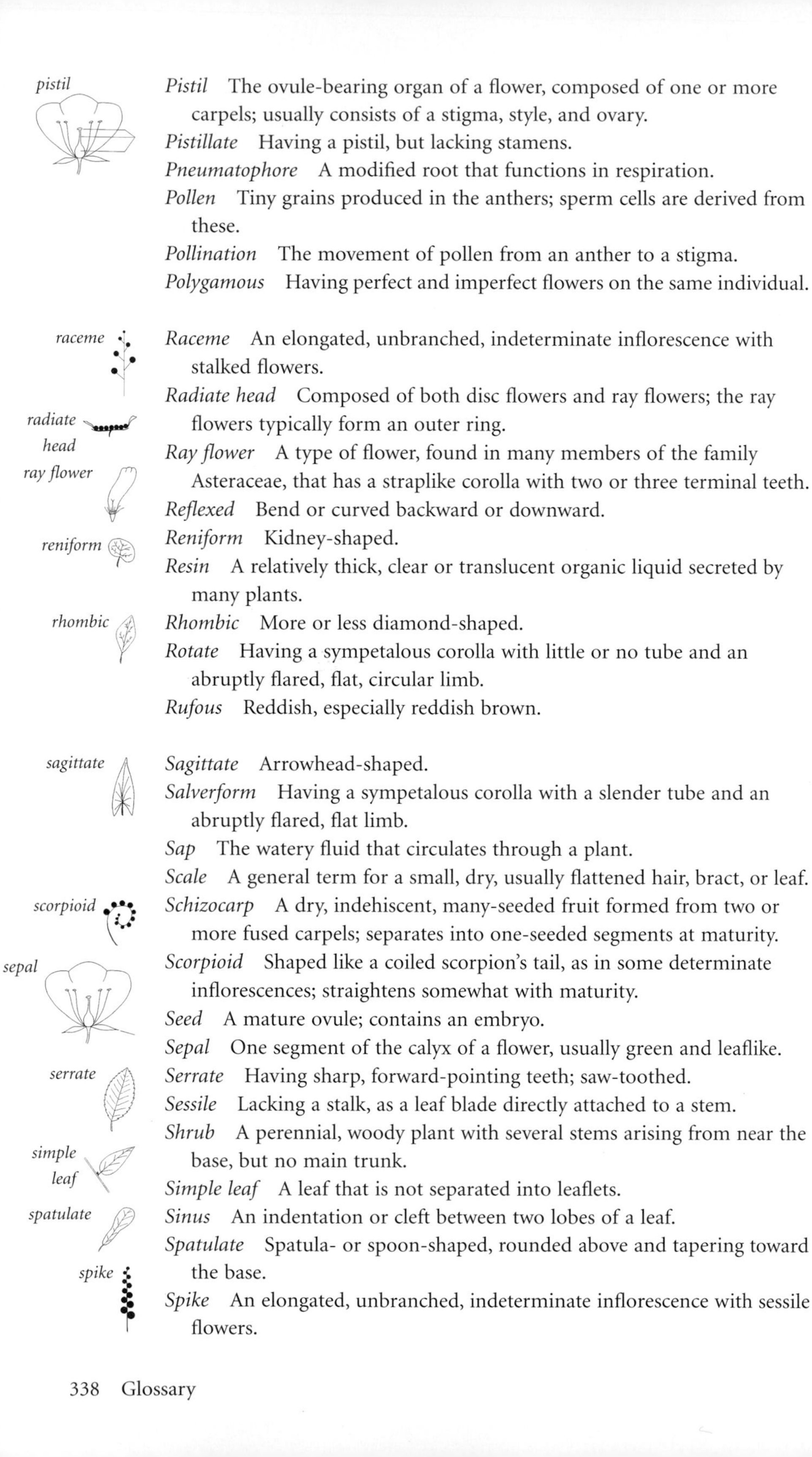

Pistil The ovule-bearing organ of a flower, composed of one or more carpels; usually consists of a stigma, style, and ovary.

Pistillate Having a pistil, but lacking stamens.

Pneumatophore A modified root that functions in respiration.

Pollen Tiny grains produced in the anthers; sperm cells are derived from these.

Pollination The movement of pollen from an anther to a stigma.

Polygamous Having perfect and imperfect flowers on the same individual.

Raceme An elongated, unbranched, indeterminate inflorescence with stalked flowers.

Radiate head Composed of both disc flowers and ray flowers; the ray flowers typically form an outer ring.

Ray flower A type of flower, found in many members of the family Asteraceae, that has a straplike corolla with two or three terminal teeth.

Reflexed Bend or curved backward or downward.

Reniform Kidney-shaped.

Resin A relatively thick, clear or translucent organic liquid secreted by many plants.

Rhombic More or less diamond-shaped.

Rotate Having a sympetalous corolla with little or no tube and an abruptly flared, flat, circular limb.

Rufous Reddish, especially reddish brown.

Sagittate Arrowhead-shaped.

Salverform Having a sympetalous corolla with a slender tube and an abruptly flared, flat limb.

Sap The watery fluid that circulates through a plant.

Scale A general term for a small, dry, usually flattened hair, bract, or leaf.

Schizocarp A dry, indehiscent, many-seeded fruit formed from two or more fused carpels; separates into one-seeded segments at maturity.

Scorpioid Shaped like a coiled scorpion's tail, as in some determinate inflorescences; straightens somewhat with maturity.

Seed A mature ovule; contains an embryo.

Sepal One segment of the calyx of a flower, usually green and leaflike.

Serrate Having sharp, forward-pointing teeth; saw-toothed.

Sessile Lacking a stalk, as a leaf blade directly attached to a stem.

Shrub A perennial, woody plant with several stems arising from near the base, but no main trunk.

Simple leaf A leaf that is not separated into leaflets.

Sinus An indentation or cleft between two lobes of a leaf.

Spatulate Spatula- or spoon-shaped, rounded above and tapering toward the base.

Spike An elongated, unbranched, indeterminate inflorescence with sessile flowers.

Spikelet The smallest flower cluster of a grass or sedge inflorescence; each holds one to many flowers and has one or more bracts directly beneath it.

Spine A stiff, sharp-pointed structure representing a leaf modification; any structure having the appearance of a true spine.

Spreading Diverging outward, almost horizontal.

Spur A hollow projection from a flower, typically formed from a modified petal or sepal, that usually collects nectar.

Spur shoot A short shoot that bears leaves or flowers.

Stamen The pollen-bearing organ of a flower, usually consisting of an anther and a filament.

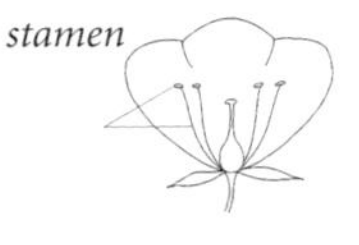

Staminate Having stamens, but lacking a pistil.

Staminodium A sterile stamen or stamenlike structure.

Standard The uppermost, typically largest petal found in the flower of many members of the Fabaceae.

Stellate Star-shaped, or branching in a star-shaped pattern.

Stigma The pistil's top part, which is receptive to pollen.

Stipular Arising from the stipules.

Stipule An appendage at the base of a petiole or leaf; often there are two of these associated with each leaf.

Style The typically narrow portion of a pistil that connects the stigma to the ovary.

style

Sympetalous Having the petals fused, at least basally.

Taxa Taxonomic groups constituting the ranks of classification, such as family, genus, and species.

Tendril An elongated, twining, modified leaf or stem that helps a plant cling to a support.

Tepal A segment of the perianth that cannot be clearly distinguished as sepal or petal.

Terminal At the apex or tip.

Thorn A sharp-pointed, often woody outgrowth of a stem; usually a reduced, modified branch.

Tooth A small point, lobe, or extension along a margin.

Tree A relatively large, perennial, woody plant that typically has a single main trunk.

Tube The constricted basal portion of certain sympetalous corollas.

Tubular Having the form of a tube, with more or less parallel sides.

Umbel An indeterminate, usually broad and more or less flat-topped inflorescence, in which the flower stalks arise from a common point.

umbel

Unisexual Having either stamens or pistil(s) in each flower, but not both; imperfect.

Utricle A dry, bladdery, usually indehiscent, one-seeded fruit formed from a single carpel.

Vascular plant A plant that possesses the conducting tissues known as "xylem" and "phloem."

Vein A strand of vascular tissue, most often observed on leaves.

Vine An herbaceous or woody plant with a long, flexible stem that trails on the ground or climbs on a support such as another plant or a wall.

Whorl A circle of three or more structures per node.

whorled

Whorled Arranged in a circle of three or more structures per node, as leaves on a stem.

Wing One of two lateral petals found in the flower of many members of the Fabaceae.

Winged Possessing a thin, flat extension on a structure.

Woolly Possessing long, soft, somewhat matted hairs.

Xylem The conducting tissue that transports water and dissolved minerals throughout a plant's body.

Literature Cited

Adsersen, H. 1980. Revision of the Galápagos endemic genus *Lecocarpus* (Asteraceae). Botanisk Tidsskrift 75: 63–76.

———. 1989. The rare plants of the Galápagos Islands and their conservation. Biological Conservation 47: 49–77.

Anderson, D. E. 1974. Taxonomy of the genus *Chloris* (Gramineae). Brigham Young University Science Bulletin 19: 1–133.

Colinvaux, P. A., and E. K. Schofield. 1976. Historical ecology of the Galápagos Islands. I. A Holocene pollen record from El Junco, Isla San Cristóbal. Journal of Ecology 64: 989–1012.

Darwin, C. R. 1845. Journal of researches into the natural history and geology of the countries visited during the voyage of H.M.S. *Beagle* round the world under the command of Capt. Fitz Roy, R.N. 2d ed. London: Murray.

———. 1859. On the origin of species by means of natural selection, or the preservation of favoured races in the struggle for life. London: Murray.

Eliasson, U. 1974. Studies in Galápagos plants. XIV. The genus *Scalesia* Arn. Opera Botanica 36: 1–117.

Elisens, W. J. 1989. Genetic variation and evolution of the Galápagos shrub snapdragon. National Geographic Research 5: 98–110.

Grant, B. R., and P. R. Grant. 1981. Exploitation of *Opuntia* cactus by birds on the Galápagos. Oecologia 49: 179–87.

Grant, P. R. 1986. Ecology and evolution of Darwin's Finches. Princeton: Princeton University Press.

Hamann, O. 1974a. Contributions to the flora and vegetation of the Galápagos Islands. I. New floristic records from the archipelago. Botaniska Notiser 127: 245–51.

———. 1974b. Contributions to the flora and vegetation of the Galápagos Islands. III. Five new floristic records. Botaniska Notiser 127: 309–16.

———. 1979. Taxonomic and floristic notes from the Galápagos Islands. Botaniska Notiser 132: 435–40.

———. 1981. Plant communities of the Galápagos Islands. Dansk Botanisk Archiv 34: 1–163.

Hamann, O., and S. Wium-Andersen. 1986. *Scalesia gordilloi* sp. nov. (Asteraceae) from the Galápagos Islands, Ecuador. Nordic Journal of Botany 6: 35–38.

Hooker, J. D. 1847. An enumeration of the plants of the Galápagos Archipelago; with descriptions of those which are new. Transactions of the Linnean Society of London 20: 163–233.

Johnson, M. P., and P. H. Raven. 1973. Species number and endemism: the Galápagos archipelago revisited. Science 179: 893–95.

Lawesson, J. E. 1988. Contributions to the flora of the Galápagos Islands, Ecuador. Phytologia 65: 228–30.

———. 1990. Threatened plant species and priority plant conservation sites in the Galápagos Islands. Monographs in Systematic Botany from the Missouri Botanical Garden 32: 153–67.

Lawesson, J. E., and H. Adsersen. 1987. Notes on the endemic genus *Darwiniothamnus* (Asteraceae) from the Galápagos Islands. Opera Botanica 92: 7–15.

Lawesson, J. E., H. Adsersen, and P. Bentley. 1987. An updated and annotated check list of the vascular plants of the Galápagos Islands. Reports from the Botanical Institute, University of Aarhus 16: 1–74.

Linsley, E .G., C. M. Rick, and S. G. Stephens. 1966. Observations on the floral relationships of the Galápagos carpenter bee. The Pan-Pacific Entomologist 42: 1–18.

Mauchamp, A. 1997. Threats from alien plant species in the Galápagos Islands. Conservation Biology 11: 260–63.

McMullen, C. K. 1985. Observations on insect visitors to flowering plants of Isla Santa Cruz. I. The endemic carpenter bee. Noticias de Galápagos 42: 24–25.

———. 1986a. Observations on insect visitors to flowering plants of Isla Santa Cruz. II. Butterflies, moths, ants, hover flies, and stilt bugs. Noticias de Galápagos 43: 21–23.

———. 1986b. Seed germination studies of selected Galápagos Islands angiosperms. Noticias de Galápagos 44: 21–24.

———. 1987. Breeding systems of selected Galápagos Islands angiosperms. American Journal of Botany 74: 1694–1705.

———. 1989. The Galápagos carpenter bee, just how important is it? Noticias de Galápagos 48: 16–18.

———. 1990. Reproductive biology of Galápagos Islands angiosperms. Monographs in Systematic Botany from the Missouri Botanical Garden 32: 35–45.

———. 1993. Flower-visiting insects of the Galápagos Islands. The Pan-Pacific Entomologist 69: 95–106.

———. 1994. Pollinator availability: a possible explanation of inter-island floral variation in *Justicia galapagana* (Acanthaceae). Noticias de Galápagos 54: 22–27.

McMullen, C. K., and D. Close. 1993. Wind pollination in the Galápagos Islands. Noticias de Galápagos 52: 12–17.

McMullen, C. K., and S. Naranjo. 1994. Pollination of *Scalesia baurii* ssp. *hopkinsii* (Asteraceae) on Pinta Island. Noticias de Galápagos 53: 25–28.

McMullen, C. K., and D. M. Viderman. 1994. Comparative studies on the pollination biology of *Darwiniothamnus tenuifolius* (Asteraceae) and *Plumbago scandens* (Plumbaginaceae) on Pinta Island and Santa Cruz Island, Galápagos. Phytologia 76: 30–38.

Meier, R. E. 1993. Coexisting patterns and foraging behavior of introduced and native ants (Hymenoptera Formicidae) in the Galápagos Islands (Ecuador). In D. F. Williams, ed., Exotic ants: biology, impact, and control of introduced species, pp. 44–62. Boulder: Westview Press.

Porter, D. M. 1980a. The vascular plants of Joseph Dalton Hooker's *An enumeration of the plants of the Galápagos Archipelago; with descriptions of those which are new.* Botanical Journal of the Linnean Society 81: 79–134.

———. 1980b. Charles Darwin's plant collections from the voyage of the *Beagle.* Journal of the Society for the Bibliography of Natural History 9: 515–25.

———. 1983. Vascular plants of the Galápagos: origins and dispersal. In R. I. Bowman, M. Berson, and A. E. Levitan, eds., Patterns of evolution in Galápagos organisms, pp. 33–96. San Francisco: American Association for the Advancement of Science.

Rick, C. M., and R. I. Bowman. 1961. Galápagos tomatoes and tortoises. Evolution 15: 407–17.

Robinson, B. L. 1902. Flora of the Galápagos Islands. Proceedings of the American Academy of Arts and Sciences 38: 78–270.

Schofield, E. K. 1984. Plants of the Galápagos Islands. New York: Universe Books.

Snell, H. M., P. A. Stone, and H. L. Snell. 1995. Geographical characteristics of the Galápagos Islands. Noticias de Galápagos 55: 18–24.

Stewart, A. 1911. A botanical survey of the Galápagos Islands. Proceedings of the California Academy of Sciences 1: 7–288.

van der Werff, H. 1977. Vascular plants from the Galápagos Islands: new records and taxonomic notes. Botaniska Notiser 130: 89–100.

———. 1979. Conservation and vegetation of the Galápagos Islands. In D. Bramwell, ed., Plants and islands, pp. 391–404. London: Academic Press.

Wagner, W. L., D. R. Herbst, and S. H. Sohmer. 1990. Manual of the flowering plants of Hawaii. Honolulu: University of Hawaii Press.

Wiggins, I. L., and D. M. Porter. 1971. Flora of the Galápagos Islands. Stanford: Stanford University Press.

Appendix 1

List of Galápagos Islands flowering plants treated in the text. An asterisk indicates that a photograph is included.

ACANTHACEAE
- *Blechum pyramidatum**
- *Elytraria imbricata**
- *Justicia galapagana**

AGAVACEAE
- *Furcraea hexapetala**

AIZOACEAE
- *Sesuvium edmonstonei**
- *S. portulacastrum**
- *Trianthema portulacastrum**

AMARANTHACEAE
- *Alternanthera echinocephala**
- *A. filifolia*
  - subsp. *filifolia**
  - subsp. *glauca*
  - subsp. *glaucescens*
  - subsp. *microcephala*
  - subsp. *nudicaulis*
  - subsp. *pintensis*
  - subsp. *rabidensis*
- *A. halimifolia**
- *Amaranthus spinosus**
- *Froelichia juncea*
  - subsp. *alata**
  - subsp. *juncea*
- *Froelichia nudicaulis*
  - subsp. *curta*
  - subsp. *lanigera*
  - subsp. *nudicaulis*
- *Lithophila radicata**
- *L. subscaposa*
- *Pleuropetalum darwinii**

AMARYLLIDACEAE
- *Crinum latifolium**

APOCYNACEAE
- *Allamanda cathartica**
- *Cascabela thevetia**
- *Catharanthus roseus**
- *Nerium oleander**
- *Vallesia glabra*
  - var. *glabra*
  - var. *pubescens**

ARECACEAE
- *Cocos nucifera**

ASCLEPIADACEAE
- *Asclepias curassavica**
- *Sarcostemma angustissimum**

ASTERACEAE
- *Adenostemma platyphyllum**
- *Ageratum conyzoides**
- *Baccharis gnidiifolia**
- *B. steetzii*

Bidens cynapiifolia
*B. pilosa**
B. riparia
*Blainvillea dichotoma**
Darwiniothamnus alternifolius
D. lancifolius
subsp. *glabriusculus**
subsp. *glandulosus*
subsp. *lancifolius*
*D. tenuifolius**
*Eclipta alba**
*Jaegeria gracilis**
*Lecocarpus darwinii**
*L. lecocarpoides**
*L. pinnatifidus**
*Macraea laricifolia**
Pectis linifolia
*P. subsquarrosa**
*P. tenuifolia**
Porophyllum ruderale
var. *macrocephalum**
Pseudelephantopus spicatus
*P. spiralis**
*Scalesia affinis**
S. aspera
S. atractyloides
var. *atractyloides*
var. *darwinii*
S. baurii
subsp. *baurii*
subsp. *hopkinsii**
*S. cordata**
*S. crockeri**
S. divisa
*S. gordilloi**
S. helleri
subsp. *helleri*
subsp. *santacruziana**
S. incisa
S. microcephala
var. *cordifolia*
var. *microcephala**
*S. pedunculata**
S. retroflexa
*S. stewartii**
*S. villosa**
*Sonchus oleraceus**

AVICENNIACEAE
*Avicennia germinans**

BATACEAE
*Batis maritima**

BIGNONIACEAE
*Spathodea campanulata**

BOMBACACEAE
*Ceiba pentandra**
*Ochroma pyramidale**

BORAGINACEAE
Cordia alliodora
C. anderssonii
*C. leucophlyctis**
*C. lutea**
C. polycephala
*C. revoluta**
C. scouleri
Heliotropium anderssonii
*H. angiospermum**
H. curassavicum
var. *curassivicum**
*H. indicum**
H. rufipilum
var. *anademum*
Tiquilia darwinii
*T. galapagoa**
*T. nesiotica**
*Tournefortia psilostachya**
*T. pubescens**
*T. rufo-sericea**

BROMELIACEAE
*Ananas comosus**
*Tillandsia insularis**

BURSERACEAE
*Bursera graveolens**
*B. malacophylla**

CACTACEAE

*Brachycereus nesioticus**
*Hylocereus undatus**
Jasminocereus thouarsii
var. *delicatus**
var. *sclerocarpus**
var. *thouarsii*
Opuntia echios
var. *barringtonensis**
var. *echios**
var. *gigantea**
var. *inermis**
var. *zacana**
O. galapageia
var. *galapageia**
var. *macrocarpa*
var. *profusa**
*O. helleri**
*O. insularis**
O. megasperma
var. *megasperma**
var. *orientalis**
var. *mesophytica*
O. saxicola

CAESALPINIACEAE

*Bauhinia monandra**
B. variegata
*Caesalpinia bonduc**
*C. pulcherrima**
*Delonix regia**
*Parkinsonia aculeata**
*Senna alata**
S. bicapsularis
var. *bicapsularis*
S. hirsuta
var. *hirsuta*
*S. obtusifolia**
*S. occidentalis**
S. pistaciifolia
var. *picta**
S. uniflora
*Tamarindus indica**

CAPPARIDACEAE

*Cleome viscosa**

CARICACEAE

*Carica papaya**

CARYOPHYLLACEAE

*Drymaria cordata**
D. monticola
D. rotundifolia

CASUARINACEAE

*Casuarina equisetifolia**

CELASTRACEAE

*Maytenus octogona**

CHENOPODIACEAE

*Atriplex peruviana**

CLUSIACEAE

Hypericum uliginosum
var. *pratense**

COMBRETACEAE

*Conocarpus erectus**
*Laguncularia racemosa**
*Terminalia catappa**

COMMELINACEAE

*Commelina diffusa**
*Tradescantia zebrina**

CONVOLVULACEAE

*Evolvulus convolvuloides**
*E. simplex**
*Ipomoea alba**
*I. habeliana**
I. imperati
*I. linearifolia**
*I. nil**
*I. pes-caprae**
*I. triloba**
*Merremia aegyptica**
*M. umbellata**

CRASSULACEAE
Kalanchoe pinnata*

CUCURBITACEAE
Cucumis dipsaceus*
C. melo
C. sativus
Cucurbita pepo
Mormordica charantia*

CYPERACEAE
Cyperus anderssonii*
C. ligularis*

EUPHORBIACEAE
Chamaesyce amplexicaulis*
C. hirta*
C. viminea*
Codiaeum variegatum*
Croton scouleri
var. brevifolius
var. darwinii
var. grandifolius
var. scouleri*
Euphorbia cyathophora*
E. equisetiformis
E. lactea*
E. milii*
E. pulcherrima
E. tirucalli*
Hippomane mancinella*
Phyllanthus acidus
P. caroliniensis
subsp. caroleniensis*
Ricinus communis*

FABACEAE
Clitoria ternatea*
Crotalaria incana
var. incana*
var. nicaraguensis
C. pumila
C. retusa*
Desmodium incanum*
D. glabrum
D. limense
D. procumbens
Erythrina corallodendron*
E. edulis
E. fusca
E. poepiggiana
E. smithiana
E. velutina*
Phaseolus adenanthus
P. atropurpureus
P. lathyroides
P. mollis*
P. vulgaris
Piscidia carthagenensis*
Rhynchosia minima*
Tephrosia decumbens*
Vigna luteola*

GOODENIACEAE
Scaevola plumieri*

HYPOXIDACEAE
Hypoxis decumbens*

IRIDACEAE
Sisyrinchium macrocephalum

LAMIACEAE
Hyptis gymnocaulos
H. mutabilis
H. rhomboidea*
H. sidaefolia
H. spicigera
Teucrium vesicarium*

LAURACEAE
Persea americana*

LOASACEAE
Mentzelia aspera*

LYTHRACEAE
Cuphea carthagenensis
C. racemosa*

MALVACEAE

*Anoda acerifolia**
*Bastardia viscosa**
*Gossypium darwinii**
G. klotzschianum
Hibiscus diversifolius
*H. rosa-sinensis**
*H. schizopetalus**
*H. tiliaceus**
Sida acuta
*S. ciliaris**
S. glutinosa
S. hederifolia
S. paniculata
*S. rhombifolia**
S. rupo
S. salviifolia
S. spinosa
S. veronicifolia

MELASTOMATACEAE

*Miconia robinsoniana**

MELIACEAE

*Cedrela odorata**
*Melia azedarach**

MENISPERMACEAE

Cissampelos glaberrima
*C. pareira**

MIMOSACEAE

*Acacia insulae-iacobi**
A. macracantha
*A. nilotica**
*A. rorudiana**
Desmanthus virgatus
  var. *depressus**
Inga edulis
*I. schimpffii**
*Leucaena leucocephala**
Mimosa spp.
*Prosopis juliflora**

MOLLUGINACEAE

Mollugo cerviana
M. crockeri
M. flavescens
  subsp. *flavescens*
  subsp. *gracillima**
  subsp. *insularis*
  subsp. *striata*
M. floriana
  subsp. *floriana*
  subsp. *gypsophiloides*
  subsp. *santacruziana*
*M. snodgrassii**

MUSACEAE

Musa acuminata
*M. x paradisiaca**

MYRTACEAE

Psidium galapageium
  var. *galapageium*
  var. *howellii**
*P. guajava**
*Syzygium jambos**
*S. malaccense**

NOLANACEAE

*Nolana galapagensis**

NYCTAGINACEAE

*Boerhaavia caribaea**
B. coccinea
B. erecta
*Bougainvillea spectabilis**
*Commicarpus tuberosus**
*Cryptocarpus pyriformis**
*Mirabilis jalapa**
*Pisonia floribunda**

ONAGRACEAE

Ludwigia erecta
*L. leptocarpa**
L. peploides
  subsp. *peploides*

ORCHIDACEAE

*Epidendrum spicatum**
Habenaria alata

H. distans
*H. monorrhiza**
*Ionopsis utricularioides**

OXALIDACEAE
*Oxalis corymbosa**
*O. corniculata**
*O. dombeyi**
O. megalorrhiza

PASSIFLORACEAE
*Passiflora colinvauxii**
*P. edulis**
*P. foetida**
*P. quadrangularis**
*P. suberosa**
P. tridactylites

PIPERACEAE
Peperomia galapagensis
var. *galapagensis**
var. *ramulosa*
P. galioides
P. obtusilimba
P. petiolata
P. tequendamana
*Pothomorphe peltata**

PLANTAGINACEAE
Plantago galapagensis
*P. major**

PLUMBAGINACEAE
Plumbago coerulea
*P. scandens**

POACEAE
Aristida divulsa
A. repens
*A. subspicata**
A. villosa
*Bambusa guadua**
Chloris mollis
C. pycnothrix
C. radiata
*C. virgata**
Eragrostis cilianensis
*E. ciliaris**
E. mexicana
E. pilosa
Pennisetum pauperum
*P. purpureum**
Sporobolus indicus
S. pyramidatus
*S. virginicus**

POLYGALACEAE
Polygala anderssonii
P. galapageia
var. *galapageia*
var. *insularis*
P. sancti-georgii
var. *oblanceolata*
var. *sancti-georgii**

POLYGONACEAE
Polygonum acuminatum
*P. galapagense**
P. hydropiperoides
var. *persicarioides*
*P. opelousanum**
P. punctatum

PORTULACACEAE
*Calandrinia galapagosa**
Portulaca grandiflora
*P. howellii**
*P. oleracea**
P. umbraticola

RHAMNACEAE
Scutia spicata
var. *pauciflora**

RHIZOPHORACEAE
*Rhizophora mangle**

ROSACEAE
Rubus bogotensis
*R. niveus**

RUBIACEAE

Borreria laevis
*Chiococca alba**
*Cinchona succirubra**
*Coffea arabica**
*Diodia radula**
Psychotria angustata
*P. rufipes**

RUTACEAE

Citrus aurantifolia
C. aurantium
*C. limetta**
C. limon
C. medica
C. paradisi
C. reticulata
*C. sinensis**
*Zanthoxylum fagara**

SAPINDACEAE

*Cardiospermum galapageium**
*C. halicacabum**
*Sapindus saponaria**

SCROPHULARIACEAE

*Calceolaria meistantha**
C. mexicana
*Capraria biflora**
*C. peruviana**
*Galvezia leucantha**
subsp. *leucantha*
subsp. *pubescens*
subsp. (undescribed)
*Russelia equisetiformis**
*Scoparia dulcis**

SIMAROUBACEAE

*Castela galapageia**

SOLANACEAE

*Acnistus ellipticus**
*Browallia americana**
*Brugmansia candida**
Capsicum annuum
*C. frutescens**
C. galapagoense
C. pendulum
*Exedeconus miersii**
*Grabowskia boerhaaviaefolia**
*Lycium minimum**
Lycopersicon cheesmanii
var. *cheesmanii**
var. *minor**
L. esculentum
Physalis angulata
P. galapagoensis
P. peruviana
*P. pubescens**
*Solandra maxima**
*Solanum americanum**
*S. erianthum**
S. melongena
*S. quitoense**
S. tuberosum

STERCULIACEAE

*Waltheria ovata**

TILIACEAE

*Triumfetta semitriloba**

ULMACEAE

*Trema micrantha**

URTICACEAE

*Pilea baurii**
P. microphylla
P. peploides
*Urera caracasana**

VERBENACEAE

Clerodendrum molle
var. *glabrescens*
var. *molle**
*C. philippinum**
*Lantana camara**
*L. peduncularis**
Lippia reptans
*L. rosmarinifolia**
var. *rosmarinifolia*
var. *latifolia*

L. salicifolia
*L. strigulosa**
*Priva lappulacea**
*Stachytarpheta cayennensis**
*Tectona grandis**
Verbena brasiliensis
V. grisea
*V. litoralis**
V. sedula
  var. *darwinii*
  var. *fournieri*
  var. *sedula*
V. townsendii

VISCACEAE
*Phoradendron henslowii**

ZYGOPHYLLACEAE
Kallstroemia adscendens
*Tribulus cistoides**
T. terrestris

Appendix 2

Selected visitor sites and flowering plants likely to be encountered.

BALTRA

Bursera malacophylla (Galápagos incense tree)
Opuntia echios var. *echios* (prickly pear cactus)
Parkinsonia aculeata (Jerusalem thorn)
Scalesia crockeri (Crocker's scalesia)

BARTOLOMÉ

Brachycereus nesioticus (lava cactus)
Chamaesyce amplexicaulis (chamaesyce)
Cryptocarpus pyriformis (salt bush)
Ipomoea pes-caprae (beach morning-glory)
Laguncularia racemosa (white mangrove)
Maytenus octogona (maytenus)
Mollugo flavescens subsp. *gracillima* (mollugo)
Opuntia galapageia var. *galapageia* (prickly pear cactus)
Pectis tenuifolia (pectis)
Rhizophora mangle (red mangrove)
Scalesia stewartii (Stewart's scalesia)
Scutia spicata var. *pauciflora* (thorn shrub)
Tiquilia nesiotica (gray matplant)

DAPHNE

Bursera malacophylla (Galápagos incense tree)
Chamaesyce amplexicaulis (chamaesyce)
Opuntia echios var. *echios* (prickly pear cactus)
Portulaca howellii (Galápagos purslane)
Tiquilia galapagoa (gray matplant)
Tribulus cistoides (puncture weed)

ESPAÑOLA

Gardner Bay

Cryptocarpus pyriformis (salt bush)
Opuntia megasperma var. *orientalis* (prickly pear cactus)
Parkinsonia aculeata (Jerusalem thorn)
Prosopis juliflora (mesquite)
Sporobolus virginicus (beach dropseed)
Tribulus cistoides (puncture weed)

Punta Suarez

Atriplex peruviana (atriplex)
Cryptocarpus pyriformis (salt bush)
Exedeconus miersii (Galápagos shore petunia)

Grabowskia boerhaaviaefolia (grabowskia)
Lantana peduncularis (Galápagos lantana)
Lycium minimum (Galápagos lycium)
Portulaca oleracea (common purslane)
Prosopis juliflora (mesquite)
Sesuvium edmonstonei (Galápagos carpetweed)
Sida salviifolia (sida)
Trianthema portulacastrum (trianthema)
Tribulus cistoides (puncture weed)

FERNANDINA

Punta Espinosa

Avicennia germinans (black mangrove)
Brachycereus nesioticus (lava cactus)
Exedeconus miersii (Galápagos shore petunia)
Laguncularia racemosa (white mangrove)
Rhizophora mangle (red mangrove)

FLOREANA

Post Office Bay

Cryptocarpus pyriformis (salt bush)
Gossypium darwinii (Darwin's cotton)
Lantana peduncularis (Galápagos lantana)
Parkinsonia aculeata (Jerusalem thorn)
Prosopis juliflora (mesquite)

Punta Cormorán

Avicennia germinans (black mangrove)
Bursera graveolens (incense tree)
Castela galapageia (castela)
Commicarpus tuberosus (wartclub)
Croton scouleri var. *scouleri* (Galápagos croton)
Cryptocarpus pyriformis (salt bush)
Eragrostis ciliaris (eragrostis)
Lantana peduncularis (Galápagos lantana)
Lecocarpus pinnatifidus (wing-fruited lecocarpus)
Maytenus octogona (maytenus)
Mentzelia aspera (stickleaf)
Nolana galapagensis (Galápagos clubleaf)
Parkinsonia aculeata (Jerusalem thorn)
Passiflora foetida (running pop)
Pectis tenuifolia (pectis)
Plumbago scandens (white leadwort)
Polygala sancti-georgii var. *sancti-georgii* (St. George's milkwort)
Prosopis juliflora (mesquite)
Scaevola plumieri (inkberry)
Scalesia villosa (longhaired scalesia)
Scutia spicata var. *pauciflora* (thorn shrub)
Vallesia glabra (pearl berry)
Waltheria ovata (waltheria)

GENOVESA

Darwin Bay

Bursera graveolens (incense tree)
Chamaesyce amplexicaulis (chamaesyce)
C. viminea (spurred chamaesyce)
Cordia lutea (yellow cordia)
Croton scouleri var. *scouleri* (Galápagos croton)
Cryptocarpus pyriformis (salt bush)
Eragrostis ciliaris (eragrostis)
Exedeconus miersii (Galápagos shore petunia)
Heliotropium angiospermum (heliotrope)
H. curassavicum var. *curassavicum* (seaside heliotrope)
Ipomoea habeliana (lava morning-glory)
Opuntia helleri (prickly pear cactus)
Rhizophora mangle (red mangrove)

GENOVESA *continued*

Prince Philip's Steps

- *Brachycereus nesioticus* (lava cactus)
- *Bursera graveolens* (incense tree)
- *Chamaesyce amplexicaulis* (chamaesyce)
- *C. viminea* (spurred chamaesyce)
- *Croton scouleri* var. *scouleri* (Galápagos croton)
- *Ipomoea habeliana* (lava morning-glory)
- *Waltheria ovata* (waltheria)

ISABELA

Elizabeth Bay

- *Avicennia germinans* (black mangrove)
- *Bursera graveolens* (incense tree)
- *Laguncularia racemosa* (white mangrove)
- *Opuntia insularis* (prickly pear cactus)
- *Rhizophora mangle* (red mangrove)

Puerto Villamil

- *Bursera graveolens* (incense tree)
- *Chamaesyce viminea* (spurred chamaesyce)
- *Cocos nucifera* (coconut palm)
- *Conocarpus erectus* (button mangrove)
- *Cordia revoluta* (revolute-leafed cordia)
- *Cucumis dipsaceus* (hedgehog gourd)
- *Furcraea hexapetala* (Cuban hemp)
- *Hippomane mancinella* (poison apple)
- *Ipomoea pes-caprae* (beach morning-glory)
- *Jasminocereus thouarsii* var. *sclerocarpus* (candelabra cactus)
- *Laguncularia racemosa* (white mangrove)
- *Opuntia echios* var. *inermis* (prickly pear cactus)
- *Parkinsonia aculeata* (Jerusalem thorn)
- *Passiflora foetida* (running pop)
- *Pectis tenuifolia* (pectis)
- *Plumbago scandens* (white leadwort)
- *Prosopis juliflora* (mesquite)
- *Rhizophora mangle* (red mangrove)
- *Sarcostemma angustissimum* (Galápagos sarcostemma)
- *Scalesia affinis* (radiate-headed scalesia)
- *Scutia spicata* var. *pauciflora* (thorn shrub)
- *Senna pistaciifolia* var. *picta* (flat-fruited senna)
- *Vallesia glabra* var. *glabra* (pearl berry)

Punta Albemarle

- *Darwiniothamnus lancifolius* (lance-leafed Darwin's shrub)
- *Hippomane mancinella* (poison apple)
- *Ipomoea pes-caprae* (beach morning-glory)
- *Jasminocereus thouarsii* var. *sclerocarpus* (candelabra cactus)
- *Nolana galapagensis* (Galápagos clubleaf)
- *Opuntia insularis* (prickly pear cactus)
- *Scalesia affinis* (radiate-headed scalesia)

Punta García

- *Bursera graveolens* (incense tree)
- *Laguncularia racemosa* (white mangrove)
- *Rhizophora mangle* (red mangrove)
- *Scalesia affinis* (radiate-headed scalesia)

Punta Moreno

- *Brachycereus nesioticus* (lava cactus)
- *Bursera graveolens* (incense tree)
- *Darwiniothamnus tenuifolius* (thin-leafed Darwin's shrub)
- *Hippomane mancinella* (poison apple)
- *Jasminocereus thouarsii* var. *sclerocarpus* (candelabra cactus)
- *Laguncularia racemosa* (white mangrove)
- *Macraea laricifolia* (macraea)

Mollugo flavescens subsp. *gracillima* (mollugo)
Sarcostemma angustissimum (Galápagos sarcostemma)
Scalesia affinis (radiate-headed scalesia)
Scutia spicata var. *pauciflora* (thorn shrub)
Sesuvium portulacastrum (common carpetweed)

Tagus Cove
Bursera graveolens (incense tree)
Castela galapageia (castela)
Chamaesyce viminea (spurred chamaesyce)
Cordia lutea (yellow cordia)
Croton scouleri var. *scouleri* (Galápagos croton)
Cryptocarpus pyriformis (salt bush)
Exedeconus miersii (Galápagos shore petunia)
Gossypium darwinii (Darwin's cotton)
Lantana peduncularis (Galápagos lantana)
Lycopersicon cheesmanii var. *minor* (hairy Galápagos tomato)
Macraea laricifolia (macraea)
Opuntia insularis (prickly pear cactus)
Scalesia affinis (radiate-headed scalesia)
Waltheria ovata (waltheria)

Urbina Bay
Bursera graveolens (incense tree)
Darwiniothamnus tenuifolius (thin-leafed Darwin's shrub)
Exedeconus miersii (Galápagos shore petunia)
Scutia spicata var. *pauciflora* (thorn shrub)

Volcán Alcedo
Baccharis gnidiifolia (baccharis)
Bursera graveolens (incense tree)
Cordia leucophlyctis (cordia)
C. revoluta (revolute-leafed cordia)
Croton scouleri var. *scouleri* (Galápagos croton)
Darwiniothamnus lancifolius (lance-leafed Darwin's shrub)
D. tenuifolius (thin-leafed Darwin's shrub)
Epidendrum spicatum (buttonhole orchid)
Heliotropium angiospermum (heliotrope)
Hyptis rhomboidea (hyptis)
Ionopsis utricularioides (ionopsis)
Ipomoea alba (moon flower)
Lantana peduncularis (Galápagos lantana)
Lippia rosmarinifolia (narrow-leafed lippia)
Macraea laricifolia (macraea)
Opuntia insularis (prickly pear cactus)
Pennisetum purpureum (elephant grass)
Peperomia galapagensis var. *galapagensis* (Galápagos peperomia)
Pisonia floribunda (Galápagos pisonia)
Sarcostemma angustissimum (Galápagos sarcostemma)
Scalesia affinis (radiate headed scalesia)
S. microcephala var. *microcephala* (small-headed scalesia)
Scutia spicata var. *pauciflora* (thorn shrub)
Tillandsia insularis (Galápagos tillandsia)
Tournefortia pubescens (white-haired tournefortia)
T. rufo-sericea (rufous-haired tournefortia)
Trema micrantha (trema)
Zanthoxylum fagara (cat's claw)

Volcán Sierra Negra
Acacia rorudiana (Galápagos acacia)
Baccharis gnidiifolia (baccharis)
Cordia leucophlyctis (cordia)

ISABELA *continued*

Croton scouleri var. *scouleri* (Galápagos croton)
Darwiniothamnus lancifolius (lance-leafed Darwin's shrub)
D. tenuifolius (thin-leafed Darwin's shrub)
Heliotropium angiospermum (heliotrope)
Hippomane mancinella (poison apple)
Jasminocereus thouarsii var. *sclerocarpus* (candelabra cactus)
Kalanchoe pinnata (air plant)
Macraea laricifolia (macraea)
Opuntia echios var. *inermis* (prickly pear cactus)
O. insularis (prickly pear cactus)
Pisonia floribunda (Galápagos pisonia)
Polygonum galapagense (Galápagos knotweed)
Sapindus saponaria (soapberry)
Scalesia affinis (radiate-headed scalesia)
S. cordata (heart-leafed scalesia)
Scoparia dulcis (sweet-broom)
Teucrium vesicarium (germander)
Tournefortia pubescens (white-haired tournefortia)
T. rufo-sericea (rufous-haired tournefortia)

PLAZA SUR

Castela galapageia (castela)
Grabowskia boerhaaviaefolia (grabowskia)
Maytenus octogona (maytenus)
Opuntia echios var. *echios* (prickly pear cactus)
Portulaca howellii (Galápagos portulaca)
Scutia spicata var. *pauciflora* (thorn shrub)
Sesuvium edmonstonei (Galápagos carpetweed)
Tribulus cistoides (puncture weed)

RÁBIDA

Avicennia germinans (black mangrove)
Bursera graveolens (incense tree)
Cordia lutea (yellow cordia)
Croton scouleri var. *scouleri* (Galápagos croton)
Cryptocarpus pyriformis (salt bush)
Evolvulus convolvuloides (purple evolvulus)
E. simplex (white evolvulus)
Galvezia leucantha subsp. *pubescens* (Galápagos shrub snapdragon)
Lycopersicon cheesmanii var. *minor* (hairy Galápagos tomato)
Maytenus octogona (maytenus)
Merremia aegyptica (hairy merremia)
Opuntia galapageia var. *profusa* (prickly pear cactus)
Sida salviifolia (sida)

SAN CRISTÓBAL

El Junco

Ludwigia leptocarpa (false loosestrife)
Miconia robinsoniana (Galápagos miconia)
Polygonum galapagense (Galápagos knotweed)
Rubus niveus (hill raspberry)

Frigatebird Hill

Bursera graveolens (incense tree)
Chiococca alba (milkberry)
Cordia lutea (yellow cordia)
Croton scouleri var. *scouleri* (Galápagos croton)
Gossypium darwinii (Darwin's cotton)
Mollugo flavescens subsp. *gracillima* (mollugo)
Piscidia carthagenensis (piscidia)
Sida ciliaris (sida)

Vallesia glabra var. *pubescens* (pearl berry)

Puerto Baquerizo Moreno

Allamanda cathartica (golden trumpet)
Blechum pyramidatum (blechum)
Clitoria ternatea (butterfly pea)
Croton scouleri var. *scouleri* (Galápagos croton)
Desmanthus virgatus var. *depressus* (slender mimosa)
Evolvulus convolvuloides (purple evolvulus)
Gossypium darwinii (Darwin's cotton)
Ipomoea nil (blue morning-glory)
Parkinsonia aculeata (Jerusalem thorn)
Passiflora foetida (running pop)
Scalesia gordilloi (Gordillo's scalesia)
Vallesia glabra var. *pubescens* (pearl berry)

Punta Pitt

Encelia hispida (Galápagos encelia)
Lantana peduncularis (Galápagos lantana)
Mentzelia aspera (stickleaf)
Nolana galapagensis (Galápagos clubleaf)
Scalesia incisa (cut-leafed scalesia)

Road from Puerto Baquerizo Moreno to El Progreso

Cardiospermum halicacabum (heartseed)
Chamaesyce hirta (chaemaesyce)
Crotalaria retusa (rattlebox)
Gossypium darwinii (Darwin's cotton)
Lantana camara (multicolored lantana)
Mormordica charantia (bitter melon)
Pennisetum purpureum (elephant grass)
Psidium galapageium var. *howellii* (Galápagos guava)
Rubus niveus (hill raspberry)
Senna obtusifolia (sicklepod)

SANTA CRUZ

CDRS and/or Puerto Ayora

Acacia rorudiana (Galápagos acacia)
Allamanda cathartica (golden trumpet)
Alternanthera echinocephala (spiny-headed chaff flower)
A. filifolia subsp. *filifolia* (thread-leafed chaff flower)
Avicennia germinans (black mangrove)
Boerhaavia caribaea (boerhaavia)
Capraria biflora (hairy capraria)
C. peruviana (smooth capraria)
Carica papaya (papaya)
Castela galapageia (castela)
Clerodendrum molle var. *molle* (glorybower)
Commicarpus tuberosus (wartclub)
Conocarpus erectus (button mangrove)
Cordia leucophlyctis (cordia)
C. lutea (yellow cordia)
Cocos nucifera (coconut palm)
Croton scouleri var. *scouleri* (Galápagos croton)
Cryptocarpus pyriformis (salt bush)
Delonix regia (flamboyant)
Heliotropium angiospermum (heliotrope)
H. curassavicum (seaside heliotrope)
Hibiscus schizopetalus (Chinese lantern)
H. tiliaceus (seaside hibiscus)
Hippomane mancinella (poison apple)
Ipomoea pes-caprae (beach morning-glory)
Jasminocereus thouarsii var. *delicatus* (candelabra cactus)
Laguncularia racemosa (white mangrove)
Lantana camara (multicolored lantana)
Maytenus octogona (maytenus)
Melia azedarach (chinaberry)

Mormordica charantia (bitter melon)
Opuntia echios var. *gigantea* (prickly pear cactus)
Parkinsonia aculeata (Jerusalem thorn)
Passiflora foetida (running pop)
Plumbago scandens (white leadwort)
Prosopis juliflora (mesquite)
Ricinus communis (castor bean)
Rhizophora mangle (red mangrove)
Scutia spicata var. *pauciflora* (thorn shrub)
Senna alata (candle senna)
Sesuvium portulacastrum (common carpetweed)
Sida ciliaris (sida)
Tournefortia psilostachya (smooth-stemmed tournefortia)
T. pubescens (white-haired tournefortia)
Tribulus cistoides (puncture weed)

Los Gemelos

Acnistus ellipticus (Galápagos acnistus)
Ageratum conyzoides (ageratum)
Alternanthera halimifolia (chaff flower)
Anoda acerifolia (anoda)
Chiococca alba (milkberry)
Darwiniothamnus tenuifolius (thin-leafed Darwin's shrub)
Diodia radula (buttonweed)
Epidendrum spicatum (buttonhole orchid)
Jaegeria gracilis (Galápagos jaegeria)
Justicia galapagana (Galápagos justicia)
Passiflora colinvauxii (Colinvaux's passion flower)
Pennisetum purpureum (elephant grass)
Peperomia galapagensis var. *galapagensis* (Galápagos peperomia)
Phoradendron henslowii (Galápagos mistletoe)
Polygonum galapagense (Galápagos knotweed)
Psychotria rufipes (white wild coffee)
Scalesia pedunculata (tree scalesia)
Sida rhombifolia (sida)
Teucrium vesicarium (germander)
Tillandsia insularis (Galápagos tillandsia)
Tournefortia rufo-sericea (rufous-haired tournefortia)
Zanthoxylum fagara (cat's claw)

Road from Puerto Ayora to Bellavista

Bastardia viscosa (bastardia)
Capsicum frutescens (bird pepper)
Carica papaya (papaya)
Clerodendrum molle var. *molle* (glorybower)
Erythrina velutina (flame tree)
Lantana camara (multicolored lantana)
Lippia strigulosa (purple lippia)
Lycopersicon cheesmanii var. *cheesmanii* (Galápagos tomato)
Mentzelia aspera (stickleaf)
Mormordica charantia (bitter melon)
Pisonia floribunda (Galápagos pisonia)
Plumbago scandens (white leadwort)
Porophyllum ruderale var. *macrocephalum* (poreleaf)
Ricinus communis (castor bean)
Tournefortia rufo-sericea (rufous-haired tournefortia)
Waltheria ovata (waltheria)
Zanthoxylum fagara (cat's claw)

Road from Bellavista to Media Luna

Acnistus ellipticus (Galápagos acnistus)
Bambusa guadua (bamboo)
Browallia americana (bush violet)
Cinchona succirubra (quinine tree)
Coffea arabica (coffee)

Cyperus anderssonii (Andersson's sedge)
Hyptis rhomboidea (hyptis)
Miconia robinsoniana (Galápagos miconia)
Oxalis corymbosa (pink wood sorrel)
Pennisetum purpureum (elephant grass)
Persea americana (avocado)
Pothomorphe peltata (pothomorphe)
Psidium guajava (common guava)
Senna occidentalis (coffee senna)
Sida rhombifolia (sida)
Syzygium malaccense (Malay apple)
Tradescantia zebrina (wandering Jew)

Santa Rosa
Adenostemma platyphyllum (adenostemma)
Brugmansia candida (angel's trumpet)
Diodia radula (buttonweed)
Epidendrum spicatum (buttonhole orchid)
Ionopsis utricularioides (ionopsis)
Rubus niveus (hill raspberry)
Syzygium malaccense (Malay apple)

Tortoise Reserve
Blechum pyramidatum (blechum)
Caesalpinia bonduc (prickly caesalpinia)
Clerodendrum molle var. *molle* (glorybower)
Heliotropium indicum (Indian heliotrope)
Hippomane mancinella (poison apple)
Pennisetum purpureum (elephant grass)
Piscidia carthagenensis (piscidia)
Pisonia floribunda (Galápagos pisonia)
Polygonum opelousanum (knotweed)
Psidium galapageium (Galápagos guava)
Psychotria rufipes (white wild coffee)
Scalesia pedunculata (tree scalesia)
Teucrium vesicarium (germander)
Tillandsia insularis (Galápagos tillandsia)
Zanthoxylum fagara (cat's claw)

Tortuga Bay
Avicennia germinans (black mangrove)
Batis maritima (saltwort)
Cryptocarpus pyriformis (salt bush)
Evolvulus convolvuloides (purple evolvulus)
Hippomane mancinella (poison apple)
Heliotropium angiospermum (heliotrope)
Heliotropium curassavicum (seaside heliotrope)
Ipomoea pes-caprae (beach morning-glory)
Laguncularia racemosa (white mangrove)
Opuntia echios var. *gigantea* (prickly pear cactus)
Prosopis juliflora (mesquite)
Rhizophora mangle (red mangrove)
Scaevola plumieri (inkberry)
Scalesia helleri subsp. *santacruziana* (Heller's scalesia)
Scutia spicata var. *pauciflora* (thorn shrub)
Sesuvium portulacastrum (common carpetweed)

Trail from Media Luna to El Puntudo
Ageratum conyzoides (ageratum)
Cinchona succirubra (quinine tree)
Cuphea racemosa (white cuphea)
Habenaria monorrhiza (fringed orchid)
Hypericum uliginosum var. *pratense* (St. John's wort)
Jaegeria gracilis (Galápagos jaegeria)
Justicia galapagana (Galápagos justicia)
Ludwigia leptocarpa (false loosestrife)
Miconia robinsoniana (Galápagos miconia)

SANTA CRUZ *continued*

Polygonum opelousanum (knotweed)
Pseudephantopus spiralis (false elephant's foot)
Stachytarpheta cayennensis (false vervain)
Vigna luteola (wild cowpea)

Trail from Puerto Ayora to Tortuga Bay

Bursera graveolens (incense tree)
Cardiospermum galapageium (Galápagos heartseed)
Cordia leucophlyctis (cordia)
C. lutea (yellow cordia)
Croton scouleri var. *scouleri* (Galápagos croton)
Erythrina velutina (flame tree)
Jasminocereus thouarsii var. *delicatus* (candelabra cactus)
Lantana peduncularis (Galápagos lantana)
Opuntia echios var. *gigantea* (prickly pear cactus)
Passiflora foetida (running pop)
Piscidia carthagenensis (piscidia)
Trema micrantha (trema)
Waltheria ovata (waltheria)
Zanthoxylum fagara (cat's claw)

Turtle Cove

Avicennia germinans (black mangrove)
Laguncularia racemosa (white mangrove)
Opuntia echios (prickly pear cactus)
Rhizophora mangle (red mangrove)

SANTA FE

Blainvillea dichotoma (blainvillea)
Bursera graveolens (incense tree)
Chloris virgata (feather fingergrass)
Cordia lutea (yellow cordia)
Cryptocarpus pyriformis (salt bush)
Lantana peduncularis (Galápagos lantana)
Maytenus octogona (maytenus)
Opuntia echios var. *barringtonensis* (prickly pear cactus)
Scutia spicata var. *pauciflora* (thorn shrub)

SANTIAGO

Buccaneer Cove

Castela galapageia (castela)
Clerodendrum molle var. *molle* (glorybower)
Opuntia galapageia var. *galapageia* (prickly pear cactus)

Espumilla Beach

Acacia rorudiana (Galápagos acacia)
Avicennia germinans (black mangrove)
Bursera graveolens (incense tree)
Clerodendrum molle var. *molle* (glorybower)
Conocarpus erectus (button mangrove)
Cryptocarpus pyriformis (salt bush)
Hippomane mancinella (poison apple)
Maytenus octogona (maytenus)
Pisonia floribunda (Galápagos pisonia)
Vallesia glabra var. *pubescens* (pearl berry)

Puerto Egas

Bursera graveolens (incense tree)
Castela galapageia (castela)
Cordia lutea (yellow cordia)
Heliotropium angiospermum (heliotrope)
Opuntia galapageia var. *galapageia* (prickly pear cactus)
Scutia spicata var. *pauciflora* (thorn shrub)
Senna obtusifolia (sicklepod)
Tribulus cistoides (puncture weed)

Sullivan Bay

Brachycereus nesioticus (lava cactus)
Bursera graveolens (incense tree)
Laguncularia racemosa (white mangrove)

Maytenus octogona (maytenus)
Mollugo crockeri (Crocker's mollugo)
M. flavescens subsp. *gracillima* (mollugo)
Scalesia stewartii (Stewart's scalesia)
Scutia spicata var. *pauciflora* (thorn shrub)
Tiquilia galapagoa (gray matplant)
T. nesiotica (gray matplant)

SEYMOUR

Bursera malacophylla (Galápagos incense tree)
Cordia lutea (yellow cordia)
Croton scouleri var. *scouleri* (Galápagos croton)
Cryptocarpus pyriformis (salt bush)
Evolvulus convolvuloides (purple evolvulus)
Maytenus octogona (maytenus)
Merremia aegyptica (hairy merremia)
Opuntia echios var. *zacana* (prickly pear cactus)
Parkinsonia aculeata (Jerusalem thorn)
Physalis pubescens (hairy ground cherry)
Portulaca oleracea (common purslane)
Scutia spicata var. *pauciflora* (thorn shrub)
Sesuvium edmonstonei (Galápagos carpetweed)
S. portulacastrum (common carpetweed)
Trianthema portulacastrum (trianthema)
Tribulus cistoides (puncture weed)

SOMBRERO CHINO

Brachycereus nesioticus (lava cactus)
Portulaca howellii (Galápagos purslane)
Sesuvium edmonstonei (Galápagos carpetweed)
Tribulus cistoides (puncture weed)

Index

All plant names, both scientific and common, are listed alphabetically. Names in italics are synonyms.